Climate Change and Its Biological Consequences

David M. Gates

University of Michigan

Sinauer Associates, Inc. • *Publishers*

SUNDERLAND, MASSACHUSETTS

THE COVER

The current climate of eastern North America is illustrated by its forest vegetation habitats (left), ranging from the tundra of the far north to the coastal
plain forests of Florida. With the expected climate warming caused by the
accumulation of greenhouse gases, over the next 50 to 80 years forest tree
species could be forced northward; if so, potential habitats could resemble
the right illustration. The models that produce such data are discussed in
Chapter 5.

Although the magnitude of the projected temperature change is not as great
as climate changes the Earth has experienced in past eons, the rate at which
these changes are occurring—over decades rather than centuries or millenia—
is virtually unprecedented and may stress the adaptive responses of many
species beyond the point where they can survive.

CLIMATE CHANGE AND ITS BIOLOGICAL CONSEQUENCES

Copyright © 1993 by Sinauer Associates, Inc.
All rights reserved. This book may not be reproduced
in whole or in part without permission from the
publisher. For information address Sinauer Associates Inc.,
Sunderland, Massachusetts 01375 U.S.A.

Library of Congress Cataloging-in-Publication Data
Gates, David Murray, 1921–
    Climate change and its biological consequences / David M. Gates.
        p.   cm.
    Includes bibliographical references and index.
    ISBN 0-87893-224-0 (pbk.)
    1. Global temperature changes—Environmental aspects—North
America.   2. Climatic changes—Environmental aspects—North America.
3. Global warming—Environmental aspects—North America.   I. Title.
QH545.T4G38   1993
574.5'222—dc20   92-36886
                                                            CIP

Printed in U.S.A.   5   4   3   2

Contents

Preface

For the first time in human history, the effluents of society are forcing the global climate to change, probably more rapidly than at almost any time in the past. Climatologists can simulate the reaction of the atmosphere and oceans to industrial emissions using mathematical models manipulated and solved by use of large computers. These global climate models—GCMs—project that the world will warm as greenhouse gases continue to build up in the atmosphere. A warmer world will have enormous impact on all plants and animals and on the sustainabilty of life as we know it.

Is it foolhardy to write a book at this time about global warming and the projected consequences? Perhaps it is. In one sense, our understanding of the subject is a moving target. New perspectives are developing rapidly. Yet the data base needed to provide convincing evidence confirming or denying global warming is slow to evolve. Global climate models are coarse in spatial resolution, restricted by cost, and limited by our understanding of cloud dynamics and the complexities of atmosphere–ocean interactions. Yet GCMs can simulate a large number of atmospheric characteristics with reasonable accuracy, including seasonal temperature and pressure patterns throughout the world and their distributions with altitude, the general air flow, the jet streams, and many other features.

Using an understanding of physical and physiological processes, forest modeling originated just over 20 years ago. It is limited by a lack of field data, a deficiency that will require years of careful measurements to correct. The modeling of animal responses to climate change is in an even more unsatisfactory state, and the modeling of plant–animal interactions with climate less satisfactory yet. But our need to understand the impact of climate on ecosystems, or on individual species, is immediate and ongoing. Bioclimatic relationships have long been of interest to geographers, ecologists, and agricultural scientists. The broad character of vegetation response to climate change is understood through various modeling efforts, but many of the detailed responses within complex ecosystems are not easily projected. A lack of information is not an adequate excuse for doing nothing. Those scientists modeling ecosystems and climate change are doing the best they can with the tools and information available. Model improvements are being made successively as new techniques and more information become available.

Originally, it was my intent to write about ecosystems of the world and their responses to climate change. It quickly became apparent that this was unmanageable. I needed to concentrate my effort where information was available and with those ecosystems I knew best. Therefore, I have focused this treatment on North America, with a few examples from Europe. The book is based on a course I taught at the University of Michigan. It begins with a discussion of the causes of climate change, the evidence for or against changes taking place, and the projections by global climate models of future climate conditions. Insight into past climates reveals how long-term climate changes have taken place and which forces in nature have brought about those changes. The pollen record, and other evidence from organisms, tells us how ecosystems may have responded to paleoclimate changes. A consideration of physiognomy and physiology gives us further understanding of the manner of plant and animal response to climate change. Three chapters recount the efforts to model the responses of forests, ecosystems, animals, and agriculture to climate warming. Finally, there is a word about the future and what might be done to mitigate the consquences of climate change.

I have told my students over and over again: "The world is a wonderful place. Go out there, understand it, and then make it better." This book is written with the hope of providing a small platform for launching some of that understanding.

I am most grateful to Allen M. Solomon for many constructive suggestions, particularly concerning the discussion of forest models. My thanks to Daniel B. Botkin for reviewing the chapter on forest models, to Philip Myers for reading the section on mammals, and to Terry Root for helping me with the section on birds.

This book is dedicated to my wife of nearly fifty years, Marian Penley Gates, who always says: "When David is writing a book, he is no fun."

DAVID M. GATES
Ann Arbor, Michigan

CHAPTER ONE

Climate Change: Cause and Evidence

THERE IS NO ISSUE of our times more challenging or more difficult to solve than the global warming caused by greenhouse gases, unless it is the explosive, exponential growth of the world's human population. The world has been warming for over one hundred years and may warm in the future at a rate unprecedented in human existence, as a direct result of industry, forest destruction, and agriculture. These activities result in the accumulation of greenhouse gases, including carbon dioxide, nitrous oxide, methane, ozone, chlorofluorocarbons, and others. These compounds, along with water vapor, are transparent to sunlight but absorb infrared heat. Their presence in the atmosphere reduces the loss of heat from the Earth's surface to outer space—the **greenhouse effect**—thereby making the world warmer.

Seldom has a scientific issue been more inherently controversial than this one, because of the difficulty of proving cause and effect (other than by inference) and because of the complexity of the mitigating measures that would be necessary in order to reduce greenhouse gas concentrations in the atmosphere. These mitigating measures are difficult scientifically, politically, economically, and in almost every other respect, but they must be tackled vigorously. Newspaper headlines have read, "The Global Warming Panic," "Feeling the Heat," "The Global Warming is Real," and "When Rivers Go Dry and Ice Caps Melt." Then the backlash headlines came along: "The Greenhouse Effect May Be Hot Air," "New Models Chill Some Predictions of Overheated Earth," "A Faulty Greenhouse?" and "Where's the Heat?". What are the facts? What do we know that could confirm or deny the existence of greenhouse warming?

1

Greenhouse gases have long lifetimes; once in the atmosphere, they remain there for decades or even centuries. This suggests that if their emission continues unabated, their concentrations will continue to grow in the atmosphere long into the future and the world will become warmer. It also suggests that we must do something soon to begin the process of mitigation.

The premise of this book is that the world is indeed warming and will soon warm at an unprecedented rate. It is also probable that strong corrective action is not likely to be taken soon enough to avoid a serious rise of the global temperature, with its consequences of rising sea levels, parched continental interiors, increased storminess, changed ocean currents, glacial melt, agricultural failures, species extinctions, and ecosystem disruptions. The natural ecosystems of forests, shrub communities, grasslands, tundra, lakes, streams, bogs, tidelands, and others, already strongly affected by human activities, will undergo changes unimagined from past experience—changes that are subtle at first but more alarming as time goes on. The longer the world waits to take action, the more serious will be the consequences and the more difficult the ultimate solution.

Evidence for Global Warming

Why do we believe that greenhouse warming of the world resulting from human activities is already under way?

The reasoning has to do with the laws of physics. We know that the greenhouse gases, by absorbing infrared heat, can produce a warmer lower atmosphere and global surface, while at the same time making the upper atmosphere (the stratosphere) cooler. In fact, the mean annual global air temperature near the Earth's surface shows a warming trend of 0.6°C over the last 130 years (Figure 1). The Southern Hemisphere record shows steady warming over most of this time, while the Northern Hemisphere record exhibits warming until the 1940s, then a cooling trend for 30 years, and a resumed warming since the 1970s.* The global temperature is a summation of the two hemispheric data bases. Seven of the ten warmest years on record for the Earth as a whole were in the 1980s through 1990. The probability of such a warming trend oc-

*Measuring the global air temperature is not exactly an easy task. In spite of the fact that approximately 700 stations all over the globe are reporting measurements, the coverage of the world is far from complete. The number of stations keeping records was less than 100 in the 1860s and grew to more than 1,700 by the 1960s, but has decreased since then. Inconsistencies have been introduced into the climate record by changes in observation schedules and methods; rapid increases in urbanization, which causes local warming; and changes in station locations. The so-called urban warming effect has been corrected out of the data shown in Figure 1. See Karl et al. (1989) for a description of these problems.

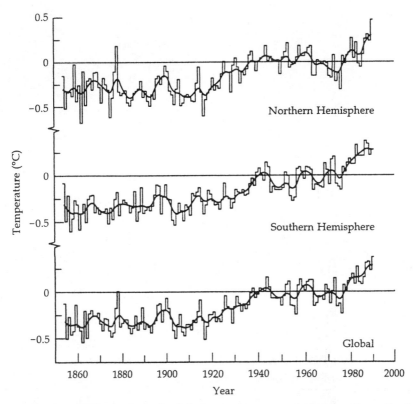

FIGURE 1 **Hemispheric and global average mean air temperatures for land plus marine regions, 1854–1990. Data are expressed as departures from the 1950–1979 average. The smooth line is a running ten-year average to show the long-term trends. (From Jones and Wigley, 1991.)**

curring only due to natural events is estimated to be less than 1% for the land and 3% for the oceans. Furthermore, the stratosphere has indeed been cooling during the last 30 years, an important manifestation of the greenhouse effect. The separate pieces of evidence for greenhouse warming are discussed later in this chapter.

The Earth's Energy Balance

Life on Earth takes place in a gaseous shell, the **atmosphere**, which includes several regions (Figure 2). Of the energy from the sun that reaches the Earth's atmosphere, part is reflected back to space by clouds and aerosols, part is absorbed selectively by atmospheric gases, and the

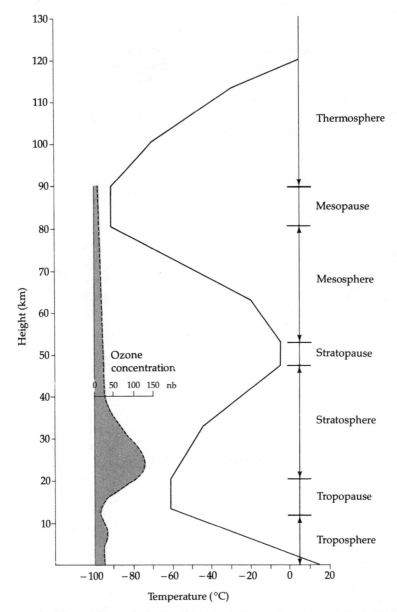

FIGURE 2   **The regions of the Earth's atmosphere, showing vertical distributions of temperature and ozone. The terms** *tropopause, stratopause,* **and** *mesopause* **apply to those boundaries where the temperature is constant with height; within each region the temperature changes with height (the** *lapse rate***). Our lives essentially occur in the troposphere; Mt. Everest, the highest point on the planet, reaches about 9 kilometers. The Concorde aircraft flies at about 10.2 kilometers.**

remainder reaches the ground, where some of it is absorbed and some is reflected upward. If these processes were all that happened, the Earth would become warmer and warmer from all the absorbed energy. But all bodies that absorb energy also radiate energy, and so does the Earth's surface. It radiates a stream of infrared heat out toward the cold of outer space. Along the way some of these infrared rays are intercepted by atmospheric gases and aerosols. Between the ground and outer space there is an ongoing exchange of some of this energy as it is absorbed, scattered, and reemitted during its outward journey.

The ground surface and lower atmosphere warm to a temperature such that the infrared energy they emit outward just balances the solar energy they absorb. If anything interferes with the amount of solar energy arriving at the ground or with the amount of infrared heat escaping to space, then the temperature of the surface will change. If the surface of the Earth were darker than it actually is, then the ground would get warmer. If high clouds or fine droplets reflect sunlight to space, then less solar energy reaches the ground and the surface becomes cooler.

Greenhouse gases have some very special properties. They are transparent to visible wavelengths of light and therefore allow sunlight to pass through the atmosphere relatively unimpeded. However, they have strong absorption regions, called bands, at infrared wavelengths, particularly around 10 μm, the wavelength of much of the Earth's radiated heat. Greenhouse gases in the atmosphere intercept and absorb some of the radiation emitted from the Earth's surface. Then they radiate this energy in all directions, some upward, some sideways, and some downward. This in effect returns energy to the ground and lower atmosphere, thus producing additional heating. A greenhouse functions in an analogous manner (Figure 3). The glass roof allows sunlight to enter the greenhouse, where it is absorbed and warms the interior; but the glass is opaque to the infrared heat, which otherwise would escape the greenhouse, allowing it to cool.

The Earth's atmosphere is truly a remarkable entity. It is mostly composed of nitrogen and oxygen gas, diatomic molecules (made up of two atoms) that are transparent to visible light, absorb some ultraviolet wavengths, and yet are transparent to infrared radiation. But these characterictistics are not shared by some of the minor constituents of the atmosphere. The triatomic molecules (three atoms) of carbon dioxide, nitrous oxide, ozone, and water vapor are each strong selective absorbers of infrared radiation, as is methane (five atoms). If the Earth's atmosphere contained only nitrogen and oxygen, the surface temperature would be about $-18°C$. However, the world has a mean surface air temperature of about 15°C because of the presence of the greenhouse gases in its atmosphere.

(A)

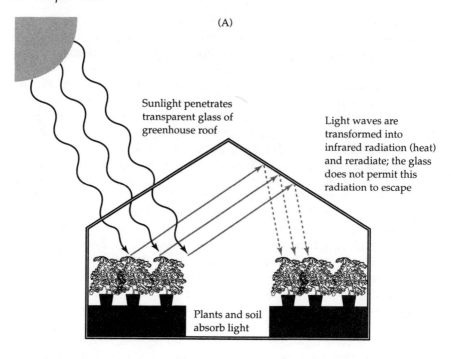

Sunlight penetrates transparent glass of greenhouse roof

Light waves are transformed into infrared radiation (heat) and reradiate; the glass does not permit this radiation to escape

Plants and soil absorb light

(B)

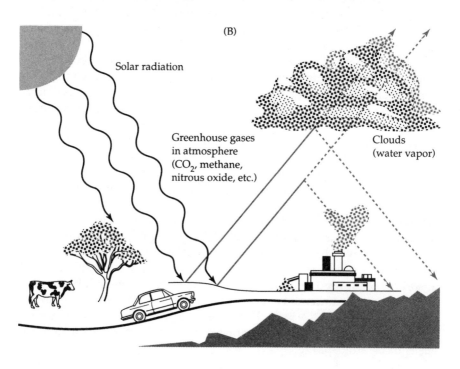

Solar radiation

Greenhouse gases in atmosphere (CO_2, methane, nitrous oxide, etc.)

Clouds (water vapor)

Nitrogen, oxygen, and ozone form a shield over the Earth that filters ultraviolet rays from sunlight, protecting life from these damaging rays. The greenhouse gases, active at infrared wavelengths, are like a blanket covering the world, keeping it warm. Between the ultraviolet filter and the infrared blanket there is a wonderful clear window to space through which sunlight streams to warm the Earth and light the flame of life through photosynthesis. Without the warm blanket our planet would be too cold for life to prosper; without the ultraviolet shield life would be destroyed; and without the clear window life would not be formed.

The energy received from the sun is not evenly distributed over the Earth; the equator is irradiated all year, the poles only half the year. The solar energy that reaches the Earth's surface is not absorbed homogeneously. Oceans, lakes, wet soil, and vegetation are quite dark and absorb sunlight strongly, while snow, ice, and dry sands reflect sunlight strongly and absorb poorly. (The measure of reflectivity is known as **albedo**, which means "whiteness.") Some of the energy that is absorbed is stored in the Earth's surface features, some of it warms the air, and the rest evaporates water from lakes, oceans, soil, and vegetation. The Earth's weather patterns and climates result from the unevenness of these processes. Figure 4 illustrates these exchanges of energy and water.

Cutting and burning forests, burning or plowing grasslands, blackening snow with dust, diverting the flow of rivers, impounding water with dams, building concrete cities and highways, and extensive agriculture—all these are modifications of the Earth's surface and they, in turn, affect the weather and climate. In many instances the changes have only local effects, but in other situations the consequences may be far-reaching in both time and space.

Greenhouse Gases

A **greenhouse gas** is any molecule that absorbs radiation in approximately the 10 μm (8–14 μm) wavelength part of the infrared spectrum—the wavelength region at which the Earth radiates heat to space. Carbon dioxide (CO_2) has long been recognized as a greenhouse gas of steadily increasing atmospheric concentration; we now realize that other rapidly increasing trace gases in the atmosphere are effective greenhouse actors

◄ FIGURE 3   The greenhouse effect. (A) In a greenhouse, the sunlight entering through the glass roof is transformed into infrared radiation (heat) that cannot escape through the glass. (B) In the atmosphere, clouds and greenhouse gases absorb some of this infrared heat and reemit it back to the ground, warming the Earth.

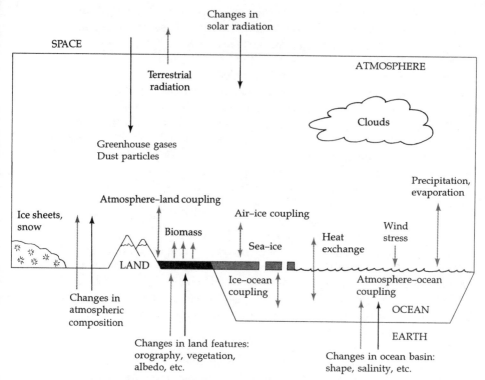

FIGURE 4 **Schematic illustration of the flow of energy and water between the land, sea, and atmosphere.**

as well. These include methane (CH_4), nitrous oxide (N_2O), ozone (O_3) in the troposphere, and the chlorofluorocarbons CFC-11 and CFC-12. Water vapor is a special case and will be considered later in this chapter for its positive feedback on the climate system. There are other gases, such as carbon monoxide (CO) and nitric oxide (NO), that are not greenhouse gases but do affect the climate through chemical reactions that generate greenhouse gases.

Table 1 shows the relative effectiveness of each greenhouse gas on a per kilogram and a per mole basis. It shows that relative to CO_2, the other gases are significantly more effective. The greenhouse effectiveness of a gas in the atmosphere depends in part on its present concentration. If a considerable amount of gas already exists in the atmosphere, as it does for CO_2, then adding more is not as effective as a whole as when the concentration of the gas is low initially. CO_2 is closer to saturation as an infrared absorber than any of the other greenhouse gases. Although CO_2 is the least effective gas on a per mole basis, it is

TABLE 1 Greenhouse gases and their contributions to global warming.

Gas	Relative effectiveness		Concentration (ppb)	Increase rate %/yr	Decay time[b] (yrs)	Current contribution (%)	Relative contribution in 100 years	
	per kg	per mole[a]					per kg	per mole[a]
CO_2	1	1	355×10^3	0.5	150–500[c]	60	1	1
CH_4	70	25	1.7×10^3	1	7–10	15–30	15–30	5–10
N_2O	200	200	310	0.2	150	5	300	300
O_3	1800	2000	10–50	0.5	0.01	8	3	4
CFC-11	4000	12,000	0.28	4	65	4	4000	11,000
CFC-12	6000	15,000	0.48	4	120	8	8000	20,000

(Data from Rodhe, 1990.)

[a]One mole of any substance contains the same number of molecules as does a mole of any other substance.

[b]The decay time is a rough measure of how long the gas remains in the atmosphere. If the decay time is, say, 150 years, one-half the initial amount remains in the atmosphere after 150 years, and one-half of that (or one-quarter of the original amount) will remain after another 150 years.

[c]The decay time for CO_2 is estimated by Rodhe (1990) as 150 years, but Wuebbles and Edmonds (1988) estimate it as 500 years.

responsible for about 60% of the greenhouse action at the present time because of its high absolute rate of increase (355 ppm $\times$ 0.5% per year increase = 1.78 ppm/year). The present concentration of each gas in the atmosphere and their current rates of increase are shown in Table 1. CO_2 is the most abundant and the CFCs the least abundant. The rates of increase range from 0.2% per year for N_2O to 4% per year for the CFCs.

Rodhe (1990) compares the effectiveness of various gas emissions by defining the accumulated greenhouse effect as the integral of the greenhouse effect over time when each gas is undergoing an exponential decrease, while at the same time being added to by continuing industrial emissions. He integrates over a 100-year period and obtains the relative contributions given in the last two columns of Table 1. The chlorofluorocarbons are the most active of the greenhouse gases, but the others are very effective as well. Fortunately, the nations of the world are beginning to phase out the production of CFC-11 and CFC-12, not only because they are active greenhouse gases but because of their threat to the stratospheric ozone concentration (Chapter 6). It is the chlorine in these molecules that catalytically destroys ozone. However, replacement compounds for CFCs that do not contain chlorine are still effective greenhouse gases.

Hansen and Lacis (1990) show that from 1850 to 1957, the contribution to climate warming of greenhouse gases other than CO_2 was only about 40% of that of CO_2, and that was primarily by methane. From 1958 to 1989 the contributions of CH_4, N_2O, and CFCs were 77% of the CO_2 contribution to warming, and during the 1990s they will equal or exceed that of CO_2.

Carbon Dioxide

Carbon dioxide gas has been increasing in atmospheric concentration since the Industrial Revolution due to the burning of fossil fuels, cement production, and forest destruction. By the 1950s the invention and commercial production of infrared gas analyzers made the accurate measurement of carbon dioxide possible on a routine basis. In 1958 a CO_2 measurement station was established at the 11,200-foot level on Mauna Loa, Hawaii. The site chosen is in the middle of the Pacific Ocean, removed from direct contamination by CO_2 emissions from large urban areas. The site is above the tropical forest of the lower slopes, where CO_2 uptake and release by plant assimilation and respiration would interfere with the measurements.

It became clear within a few years that the global CO_2 concentration was increasing inexorably and that there was an annual pulse in the record (Figure 5). This annual pulse reflects the metabolism of plants

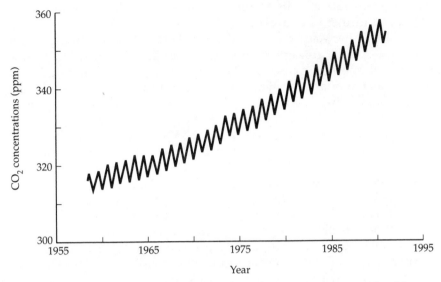

FIGURE 5 The atmospheric carbon dioxide concentration measured at Mauna Loa, Hawaii from 1958 to 1991. (Courtesy of C. D. Keeling.)

assimilating CO_2 during the growing season and releasing CO_2 during the winter. Another well-calibrated measurement station was established in 1958 at the South Pole. The amount of CO_2 recorded there was precisely the same as at Mauna Loa, but the annual pulse was very small. By the time air masses mixed around the world and reached the South Pole, the vegetation assimilation and respiration cycle was mostly damped out. Data from other stations established later at higher latitudes in the Northern Hemisphere (such as Point Barrow, Alaska and the northwest Pacific Ocean) showed larger annual variations but agreed with the steady upward trend of the annual average CO_2 concentration. The Northern Hemisphere records showed greater vegetation influence than records from much of the Southern Hemisphere, since the Northern Hemisphere has more land area and therefore more annual assimilation and respiration by plants.

The preindustrial level of atmospheric CO_2 has been measured from air trapped in ice cores in the Greenland and Antarctic ice caps. Neftel et al. (1982) give a value of about 271 ppm for 350 to 500 years ago. Liss and Crane (1983) review other data and give values between 280 and 290 ppm. However, if one projects all modern measurements made after 1860 backward in time, one gets a preindustrial value of about 275 ppm. During the last Ice Age, 40,000 years ago, atmospheric CO_2 concentrations were as low as 180 ppm, according to Chappellaz et al.

(1990). (This fact and its relationship to methane concentrations are discussed further in Chapter 2.)

In 1958 the Mauna Loa and South Pole measurements gave an atmospheric CO_2 concentration of 315 ppm; in 1991 it was 355 ppm—an annual rate of increase of 1.78 ppm per year. What the future will bring is an important question. One uncertainty about future projections of CO_2 concentrations is what growth rate to assume for the use of fossil fuels. The rate of increase during the period before 1973 was 4.3% per year; since then it has been about 2.25% per year. If we assume a 2% per year growth rate in the use of fossil fuels and an airborne fraction of 0.55, then the preindustrial level of CO_2 will double to 550 ppm by about 2053. (The *airborne fraction* is that part of the total emissions which remains in the atmosphere. The remainder goes to various sinks such as the ocean and the biosphere.) If the people of the world immediately limited the growth rate of the use of fossil fuels to zero, then doubling would not occur until late in the twenty-second century. However, there is no prospect of that occurring. At a 4% per year annual growth rate, a doubling of the preindustrial level would occur by the year 2032, again with an airborne fraction of 0.55. But we must remind ourselves that carbon dioxide is not the only greenhouse gas and that the others, except for N_2O, have even greater annual rates of increase.

The carbon budget for the atmosphere has a number of uncertainties, particularly with respect to the sinks for carbon. The carbon cycle is illustrated in Figure 6. Reservoir amounts of carbon are given in gigatons (Gt; 1 Gt = 1 billion metric tons) and annual fluxes in Gt per year. The input of carbon to the atmosphere from fossil fuel burning is best established at about 5.3 Gt per year. Of this amount, about 3.0 Gt per year remains in the atmosphere. The remainder, 2.3 Gt per year, is apparently taken up by the oceans and forests. The ocean takes up about 92 Gt per year and releases about 90 Gt per year back to the atmosphere through natural biological and chemical reaction fluxes. The terrestrial biosphere has assimilation and respiration fluxes of about 102 and 100 Gt per year, respectively. Deforestation of tropical forests releases approximately 2 Gt of carbon per year, although this amount is quite uncertain. It is also possible that the oxidation of carbon monoxide entering the atmosphere adds an additional 1 Gt per year. Thus the carbon budget is not balanced.

Tans et al. (1990), in an elaborate analysis using the global network of CO_2 concentration data and global climate modeling, come to the conclusion that the oceans cannot provide a sufficient sink for excess atmospheric CO_2, particularly in the Northern Hemisphere. They suggest that a large amount of CO_2 is absorbed on the continents by terrestrial ecosystems and that these sinks are larger in the Northern

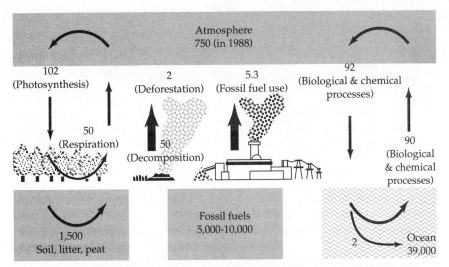

FIGURE 6 **The generalized carbon cycle for the Earth, showing the main reservoirs (in gigatons, or Gt) and fluxes (in Gt per year). Detailed reviews are given by Tans et al. (1990) and by Liss and Crane (1983).**

Hemisphere than they are in the Southern Hemisphere. These sinks may be on the order 2.0 to 3.4 Gt per year, depending on the amount of carbon released by sources in the tropics and in boreal and tundra regions. The implication is that these sinks are terrestrial and exist at temperate latitudes. Beyond this, there is a potential problem for the future. As the world warms, it is expected that soil respiration will increase, releasing more carbon, and this could be a significant positive feedback to the greenhouse effect.

The reservoirs of carbon in the world are massive. Sediments contain an estimated 20,000,000 Gt, the oceans 38,400 Gt, the terrestrial biosphere 1,760 Gt, and the fossil fuels, including shales, 12,000 Gt, of which 7,500 Gt is thought to be recoverable. It is clear that there is potential for the release of enormous additional amounts of carbon dioxide to the atmosphere.

Methane

Methane is a more effective greenhouse gas by 70 times per kilogram and 25 times per mole than carbon dioxide, and on a 100-year basis it is 5 to 10 times more effective per mole (Table 1). Its decay time is 7 to 10 years, which accounts for its reduced effectiveness as a greenhouse gas over a long time period. However, its rate of increase in the atmosphere is about 1% per year, twice the rate of increase of CO_2. Its current concentration is 1.7 ppm (1700 ppb). However, because it reacts photochemically in the atmosphere to form O_3, CO_2, and water vapor, its indirect contribution to greenhouse warming is estimated to be an additional 15%.

Crutzen (1991) describes the action of methane well:

Methane's influence is wide ranging in atmospheric chemistry. In the background troposphere, methane is involved in photochemical reactions that determine the concentrations of ozone (O_3) and hydroxyl (OH), the 'detergent' of the atmosphere, responsible for the removal of almost all gases that are produced by natural processes and human activities. Methane also influences the chemistry of the stratosphere: its oxidation is an important source of stratospheric water vapour (H_2O) and thus directly of OH radicals, whose reactions lead to the conversion of ozone-destroying NO and N_2O catalysts to far less reactive nitric acid (HNO_3). On the other hand, reactions of OH radicals with hydrochloric acid (HCl), promote the formation of ozone-destroying Cl and ClO radicals.

Measurements of methane concentrations from air trapped in the ice at Vostok, Antarctica, showed amounts over the last 160,000 years to range from as low as 350 ppb during full glacial conditions to 650 ppb during interglacial times (see Chapter 2). Beginning between 200 and 300 years ago, methane concentration started increasing, from 700 ppb to its present concentration of 1,700 ppb. The inescapable conclusion is that the increase has been caused by human activities. Methane is produced by the anaerobic decomposition of organic matter in wetlands, rice fields, landfills, the guts of termites, and the rumens of cattle. Human activities such as coal mining; oil and gas extraction, refining, and distribution; and biomass burning also generate methane. Methane emissions are growing with the increasing human population, but eventually there will be a space limitation to some of these sources. The doubling time for methane in the atmosphere is about 46 years, or twice the rate for CO_2 or for fossil fuel use.

Table 2 shows a methane budget, with rough estimates for principal biogenic sources, including ruminants, landfills, termites, and tropical sources (Crutzen, 1991). The difference between the amount of methane produced by biogenic sources and the amount of methane in atmospheric sinks is about 100 Tg/yr; this additional methane is attributed to the fossil fuel economy, including gas and oil leaks. Papers by Craig et al. (1988) and Wahlen et al. (1989) give further details on the methane budget.

Large amounts of methane are stored in methane *clathrates*, or *hydrates*, sequestered in permafrost at considerable depths onshore and offshore in polar regions. Methane clathrates are lattice-like structures

TABLE 2　A methane budget.

Methane source	Amount of methane produced (Tg/yr)[a]	Methane sink	Amount of methane in sink (Tg/yr)[a]
Methane from biotic sources:		*Fixed methane:*	
Ruminants	80	Reaction with OH	420
Landfills	50	Soil uptake	30
Ocean production, termites	30	Stratospheric	
Tropical:		reactions	10
Wetland ⎫		SUBTOTAL	460
Rice production ⎬	245		
Burning ⎭		Methane in	
SUBTOTAL	405	atmosphere	45
Fossil fuel methane production	100[b]		
Total methane from all sources	505	Total methane into sinks	505

(Data from Crutzen, 1991.)
[a]Amounts are in teragrams; 1 Tg = 1 trillion (10^{12}) grams = 1 million (10^6) metric tons.
[b]At least 70 Tg/yr of this methane is from leaks in natural gas distribution systems and from gas and oil wells.

in which methane gas is trapped in a matrix of water molecules. Clathrates are stable at low temperatures and relatively high pressures; however, if deep-soil or ocean-bottom temperatures rise, it is possible they could become unstable and release methane.

Nitrous Oxide

Nitrous oxide, N_2O, is 200 times as effective as CO_2, on either a mass or mole basis, as a greenhouse gas (Table 1). It remains in the atmosphere for about as long as CO_2 and is increasing at 0.2% per year. Most atmospheric N_2O comes from the natural nitrogen cycle of the biosphere, in which fungi and bacteria act in the denitrification of compounds in the soil. The source of the small increase in the atmosphere is not well established, but it could relate to the increasing use of fertilizers and the high combustion temperatures of fossil fuel-burning power plants and engines. N_2O is likely to continue to increase in the future because of the continuing demand of the every-growing human population for food and energy.

Ozone

Ozone (O_3) is a powerful infrared absorber and therefore ostensibly a very effective greenhouse gas. Ozone occurs naturally in the stratosphere, where it is generated photochemically by sunlight acting on oxygen molecules. Unfortunately, this stratospheric ozone, which is an essential component of the Earth's ultraviolet shield, is currently being destroyed by the catalytic action of chlorine molecules from chlorofluorocarbons released by human activity (Chapter 6). On the other hand, ozone is increasing in the lower atmosphere (the troposphere) as a result of air pollution from combustion of fossil fuels. Ozone has a rapid decay time in the troposphere. As a greenhouse gas it is only about four times as effective as CO_2 on a 100-year basis. Its rate of increase in the atmosphere is about the same as that of CO_2. At the present time it is believed to be contributing about 8% of the total greenhouse effect. Ozone has no significant direct human sources, but in the troposphere it is produced indirectly through chemical reactions involving emissions of methane, carbon monoxide, oxides of nitrogen, and hydrocarbons. Although the ozone concentration in the troposphere is generally only about 10 to 50 ppb, it can be much higher locally due to air pollution.

Chlorofluorocarbons

The chlorofluorocarbons CFC-11 and CFC-12, also known as freons, came into widespread use during the 1960s for air conditioning, as cleansing agents for electronic circuitry, and as foaming agents for the manufacture of plastic foam products. At first they grew in use at a rate of 10–15% per year, but this increase has slowed to 4% per year. They are extremely effective greenhouse gases because they possess infrared absorption bands centered in the 10-μm window. Furthermore, they have slow decay rates; stratospheric photolysis is the only means by which they are destroyed in the atmosphere. On a 100-year basis they are by all odds the most effective of the greenhouse gases. Fortunately, there are now international agreements to phase out the use of some of these compounds.

Climate Forcing Factors and Feedbacks

Hansen et al. (1990) stated that "a climate forcing is a change imposed on the planetary energy balance that alters global temperature." Examples of forcing factors are changes in solar irradiance, stratospheric aerosol particles (particularly from large volcanos), tropospheric aerosol particles (primarily from air pollution, but also from dust and vegetation), cloud cover (both natural clouds and those induced by human-generated aerosols and aircraft contrails), modifications of surface features, and, finally, greenhouse gases. Climate response to a forcing depends very much on how long the forcing is sustained. The response time of the climate system is generally quite long, often being weeks to years and sometimes decades or even centuries.

Solar Irradiance

Some scientists have long believed that sunspot cycles affect the Earth's climate. It would be surprising if solar activity had no influence on climate, considering how dramatic are some of the events on Earth associated with solar activity. Among such events are aurorae, magnetic field distortions, communication disruptions, and even electric power transmission breakdowns. Sun–Earth orbital changes have apparently affected climate over long periods of time (as described in Chapter 2), and therefore it is reasonable to expect that solar activity (sunspots, faculae, prominences, and solar storms) may have climate consequences. However, *proving* that they do has been anything but easy.

There is circumstantial evidence that the climate responds to solar activity. For example, during the seventeenth century a lull in solar activity, the Maunder Minimum, was accompanied by a cooling of the climate, an event known as the Little Ice Age (Chapter 2). Labitzke and Van Loon (1988) showed that solar activity changes influence the circulation and temperature of the troposphere and stratosphere. The influence depends on the direction of the wind in the tropical stratosphere, which reverses itself from easterly to westerly and back again about every two years, a phenomenon termed the quasibiennial oscillation.

Although sunspots appear dark against the bright solar surface (Figure 7), they are associated with an increase in the number of faculae, solar flares, and prominences; the result is increased emission of solar radiation. At such times there is also an increase in emission of ionizing radiation, including X rays, electrons, protons, and ions. The bombardment of the Earth's atmosphere by some of these energetic particles produces the aurorae, magnetic field disturbances, and other phenomena referred to above. The number of sunspots and their positions on the solar surface vary with time. Sunspots originate at high solar latitudes and migrate toward the sun's equator in what is called a "butterfly" pattern. Sunspots near the solar equator are more likely to irradiate the Earth than are those at high latitudes. Sunspot numbers reach a maximum every 11 years. Magnetic polarity of sunspot pairs reverses every 22 years. There also appears to be a longer-term modulation of solar activity, an 85-year periodicity known as the Gleissberg Cycle.

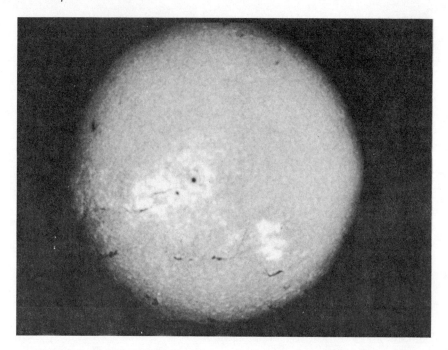

FIGURE 7   In this photograph of the sun, sunspots show up as dark patches near the sun's equator. The bright areas surrounding the sunspots are known as *plage*.

Satellite-borne radiometers measuring solar irradiance have been in operation in space since 1980. Between the sunspot maximum in 1981 and minimum in 1986, total irradiance declined about 0.1%, but since then it has increased. Global climate models show that the mean global air temperature would increase only about 0.02°C with a 0.1% irradiance increase, corresponding to 1.1 watts of energy per square meter (W/m^2). Here is an example of a forcing that is too slight to have a large impact on climate. The heat storage capacity of the oceans damps out a larger temperature change. In fact, ocean surface temperature changes lag behind land air temperature changes by several years. There has been a small rise of solar irradiance over the last 45 years, and this has produced an estimated air temperature increase of about 0.02°C (see Foukal and Lean, 1990).

Global climate models indicate that a 2% increase in solar irradiance at all wavelengths would be likely to produce approximately the same climate change (temperature increase) as a doubled concentration of carbon dioxide for the same length of time. However, there would be differences. Solar heating is concentrated in low latitudes (near the equator), whereas greenhouse heating, although more uniformly distributed, affects high latitudes more than low latitudes. Furthermore, solar radiation is most variable at ultraviolet wavelengths, and most ultraviolet radiation is absorbed high in the Earth's atmosphere. If solar irradiance continues to increase with future sunspot cycles, as it has during the last 45 years, then this will exacerbate the greenhouse warming. But if irradiance begins to decrease, as Seitz et al. (1989) suggest, it may provide some very small compensation for greenhouse warming.

An analysis by Friis-Christensen and Lassen (1991) shows a dramatic synchrony of the length of the solar cycle with variations of the Northern Hemisphere air temperature over land during the past 130 years, as shown in Figure 8. For some reason not yet understood, the length of the solar cycle seems to reflect the intensity of solar activity. Just how the climate is affected by solar activity is not yet clear, but the correlation is strong. All 11-year solar cycles are not of exactly the same duration: some are as short as 9.7 years and others as long as 11.8 years. High solar activity goes with short solar cycles, whereas long solar cycles have low activity levels. Sunspot numbers per se may not be the best indicator of solar activity. For example, high sunspot numbers have not

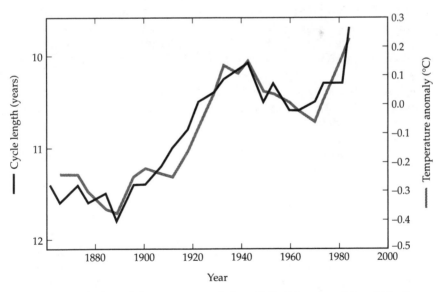

FIGURE 8 **Variation of the sunspot cycle length (left-hand scale) and Northern Hemisphere temperature anomalies from 1861 to 1989. (From Friis-Christensen and Lassen, 1991.)**

always resulted in frequent and strong aurorae; the interaction between the solar wind (ionized gas flowing into interplanetary space and bombarding the Earth's atmosphere) and the geomagnetic field that results in aurorae is not always related to sunspot numbers. Furthermore, the length of the sunspot cycle stretches and shrinks with a periodicity of 80–90 years (the Gleissberg Cycle). Friis-Christensen and Lassen show that each maximum in this long-term cycle of solar activity has been associated with minimum sea ice around Greenland, in 1770, 1850, and 1940. The match between sunspot cycle length and land air temperatures shown in Figure 8 is equally valid if a monotonically increasing greenhouse effect is superimposed on this 130-year record. Solar irradiance may have varied by approximately 0.6% (6.6 W/m^2) between 1910 and 1960. Such a long-term variation may have been considerably more effective in producing a climate change than have the shorter-term 11-year sunspot cycles.

Volcanos and Aerosols

Aerosols—suspended particles and droplets in the atmosphere—may produce a temperature reduction by reflecting sunlight back into space. Aerosols in the troposphere are likely to be washed out by precipitation or fall out by dry deposition and have short lifetimes (days or weeks). Aerosols in the stratosphere, however, may persist for several years. Fossil fuel burning and agricultural dust produce tropospheric aerosols. Many volcanos do likewise. However, the very largest volcanic eruptions will inject sulfur, water vapor, and ash into the stratosphere. In addition, high-flying jet aircraft and rockets may leave debris in the stratosphere.

That sulfuric acid aerosols injected into the stratosphere reflect sunlight is evident from many observations. Mass and Portman (1989) report that the direct solar beam may be reduced 20% to 30% by stratospheric aerosols, but since most of that radiation is scattered forward to the Earth's surface, this results in a relatively small decrease in total global irradiation of 5% to 7%.

Benjamin Franklin noted that volcanic aerosols reflect sunlight and suggested that the eruption of a large Icelandic volcano may have caused the unusual cold of 1783–1784 in Europe. Franklin at the time was ambassador to France and noticed a haze over Europe (Mitchell, 1982). The largest and deadliest volcanic eruption in recorded history was the explosion of Mount Tambora on the island of Sumbawa, Indonesia on 10 and 11 April of 1815. The eruption column probably permeated the stratosphere with ash and sulfur. Brilliantly colored sunsets were seen in London during the months following the eruption. In Europe and America, 1816 became known as "the year without a summer." Strothers (1984) performed a convincing analysis to show that Tambora most likely caused global temperature reductions. A fascinating account of the Tambora event and the climate consequences is given by Stommel and Stommel (1983).

Lough and Fritts (1987) used tree-ring data to study the effects of volcanic eruptions on surface temperatures over the United States and Canada. They showed that major low-latitude eruptions may have produced summer cooling to the east of the Rocky Mountains. Catchpole and Hanuta (1989) reconstructed the summer sea ice severity in Hudson Strait and Hudson Bay during the eighteenth and nineteenth centuries from the logbooks of sailing ships. They found that the years with the greatest sea ice severity were preceded by major volcanic eruptions. This correlation was 99.5% for ice in Hudson Strait and 95% for ice in the western part of Hudson Bay. The years with severe ice in eastern Hudson Bay did not correlate with major eruptions. Lamb (1970) proposed that volcanic dust may change the north–south gradient of temperatures, causing changes in the intensity of the atmospheric air circulation, which alters the long-wave structure of the upper atmosphere westerly winds. Arctic air frequently surges southward on the western side of a trough formed east of the Rocky Mountains. This brings cold arctic air over the western part of Hudson Bay. Exactly how the trough is nudged eastward by volcanic aerosols in the arctic stratosphere is still open to debate.

An analysis of mean monthly global air temperatures done by Kelly and Sear (1984) and corrected by Mass and Portman (1989) show a definite volcano–climate relationship. Only the most massive volcanos are effective at modifying the climate. The maximum volcano-induced surface temperature decrease was found to have been 0.34°C over land and 0.14°C over water. The temperature reduction seems to occur in the second and third years following an eruption and then disappears.

Several features characterize volcanic effects on climate. First, the volcano must be sufficiently large to inject material into the stratosphere. Second, much of that material must be small droplets of sulfuric acid, which form from volcanic sulfur dioxide. Third, several months to a year are needed for the volcanic aerosols to spread beyond the latitude band where they were injected. Fourth, surface air temperatures, even of land masses, are affected by those of the oceans, and the ocean temperature cannot respond to external forcing over a period of only a few months; hence the delay into the second year. Fifth, volcanic forcing of climate is irregular and extremely transitory. Therefore, it takes a clustering of several volcanic eruptions to have a significant impact on long-term climate. Jacoby et al. (1988) report on a major cooling period in high-latitude North America that occurred early in the nineteenth century and lasted about three decades; they attribute it in part to volcanic activity.

On 15 June 1991, the largest volcanic eruption of this century, that of Mount Pinatubo, occurred in the Philippines. This volcano sent massive amounts of sulfur dioxide gas and dust into the stratosphere to heights of more than 25 kilometers. The amount of sulfur was estimated from satellite-borne instrumentation to be between 15 and 20 million metric tons—more than twice the estimate for the eruption of El Chichón in Mexico on 4 April 1982. Scientists have been able to track the stratospheric gas cloud as it encircled the globe and spread north and south of the equator. By September 1991, Japanese scientists reported sunlight over Japan to be reduced by 15%. Sunrises and sunsets were distinctly more brilliant throughout the world during 1991 and 1992. The summer of 1992 in the northeast and upper midwest United States was by far the coldest of many decades—truly a "Pinatubo summer." Pinatubo has given climatologists a magnificent natural climate experiment. The El Chichón eruption cooled the Earth's surface air temperature by a few tenths of a degree centigrade. Global climate models by J. Hansen and colleagues at NASA Goddard, New York, show that by late 1992 the Pinatubo gas cloud may depress the global mean temperature by 0.5°C to 1.0°C, exceeding the accumulated forcing of climate by all the greenhouse gases added to the atmosphere since the Industrial Revolution. This cooling should last only a couple of years as the cloud slowly dissipates in the stratosphere. By coincidence, both the El Chichón and Pinatubo eruptions have occurred in El Niño years, in which warm Pacific Ocean waters warm the climate in both North and South America, and to some extent elsewhere (see Chapter 7). In fact, February 1992 in the United States was the warmest February on record. After El Niño has gone away and the Pinatubo cloud has diminished, greenhouse warming and solar effects should continue as the dominant forcing functions.

Atmospheric Pollution

Tropospheric aerosols may also be an important climate forcing factor. The temperature record for the Northern Hemisphere (Figure 1) exhibits a cooling trend from the 1940s to the 1970s. This contrasts with the Southern Hemisphere, where the mean annual temperature record does not exhibit the cooling but has an inexorable slow upward trend. This has led to speculation as to why the Southern Hemisphere should not have gone through the same cooling for three decades as did the Northern Hemisphere. Although the two hemispheres differ greatly in their proportions of land mass versus oceans, this disparity might account for only some of the temperature response differences. A more significant factor from the climate standpoint seems to be the much higher density of industrialization in the Northern Hemisphere, and hence the greater air pollution—a post-World War II evolution.

The mechanized agriculture of the Northern Hemisphere puts dust into the air; the slash-and-burn agriculture widely used in the Southern Hemisphere puts smoke aloft. It is generally assumed that much of this lower atmosphere contamination settles out of the air fairly quickly, say within a few days. Also, much of this agricultural contamination is seasonal and discontinuous. On the other hand, the emission of sufur dioxide (SO_2) from fossil fuel combustion and metal smelting in the industrialized nations has increased enormously. The natural global flux of SO_2 is between 20 and 30 million metric tons per year; Northern Hemisphere natural emissions are in the range of 10 to 18 million metric tons per year. Following World War II, the anthropogenic emission of SO_2 rose from about 40 million to nearly 100 million metric tons per year by the 1980s. Because residence times of SO_2 and sulfate in the troposphere are relatively short (days) compared with air mass mixing times throughout the hemisphere (months), or between hemispheres (one year), anthropogenic sulfate is concentrated mainly in industrialized regions and downwind from them. Some widespread distribution of these sulfate compounds is indicated by their presence in precipitation at remote sites in the Northern Hemisphere but not in the Southern Hemisphere. Northern Hemisphere glacial ice cores reveal concentrations of sulfate consistent with the emission rates mentioned above.

The sulfate compounds in the atmosphere, including sulfuric acid, form aerosols containing droplets of submicrometer size, which contribute substantially to the scattering of shortwave sunlight; this scattering in effect reflects some of the incident sunlight back to space. These anthropogenic compounds modify the heat balance of the Earth differently than greenhouse gases do. Aerosol climate forcing has its greatest effects in daytime and in summer, whereas greenhouse gas forcing acts over the full diurnal and seasonal cycles. Different aerosol particle sizes affect different radiation wavelengths, and aerosol forcing affects mainly the visible and near-infrared wavelengths, whereas the greenhouse gases act at longer infrared wavelengths. Maximum scattering of sunlight occurs at wavelengths approximately equal to the particle size.

Aerosols have a direct effect on climate by scattering and absorbing incoming solar radiation, and an indirect effect through their influence on the radiative properties of clouds. Clouds form by the condensation of water on cloud condensation nuclei (CCN). Sulfate aerosols are an excellent source of CCN, and these increased concentrations of CCN result in increased concentrations of cloud droplets and increased reflection of sunlight by clouds. According to Charlson et al. (1992):

> A decrease in mean droplet size associated with an increase in cloud-droplet concentration is also expected to inhibit precipitation development and hence to increase cloud lifetimes. Such an enhancement of cloud lifetime and the resultant increase in fractional cloud cover would increase both the short- and longwave radiative influence of clouds. However, because this effect would predominantly influence low clouds, for which the shortwave influence dominates, the net effect would be one of a further cooling. Inhibited precipitation development might further alter the amount and vertical distribution of water and heat in the atmosphere and thereby modify the Earth's hydrological cycle.

Charlson and colleagues estimate that anthropogenic sulfates produce an effect on the Northern Hemisphere climate comparable in magnitude but opposite in direction to the influence of all accumulated greenhouse gases. The direct aerosol and indirect cloud components contribute about equally to the climate effect.

Dimethylsulfide

In the oceans there is a different source of cloud condensation nuclei. It is dimethylsulfide (DMS), which is produced by plankton in the upper 50 meters or so of the oceans and then oxidized in the atmosphere to form sulfate aerosols. Gas exchange between this upper layer of the ocean and the atmosphere is on the order of days to weeks. Andreae (1986) estimates that the input of DMS from the oceans into the atmosphere is about one-third to one-half the amount of sulfur that enters the atmosphere from fossil fuel burning. The input from fossil fuel combustion is concentrated in the industrial centers of the Northern Hemisphere, whereas the input from DMS is spread uniformly over the world. Charlson et al. (1987) report oceanic fluxes of dimethylsulfide and show how this along with carbon dioxide fluxes can lead to stabilizing feedbacks in the ocean–atmosphere–life system. It was this idea of homeostasis that led Lovelock (1979) to propose the Gaia hypothesis.*

Radiative Heating

The most meaningful way to compare the effects of solar activity, volcanic eruptions, anthropogenic aerosols, and greenhouse gases on climate is to look at the forcings in terms of the Earth's energy budget. A good starting point for this is to consider the total amount of solar

---

*Lovelock hypothesized a system (named for Gaia, the ancient Greek concept of the Earth Mother) in which the physical and chemical environments interact with all life forms on Earth to form a single self-regulating entity with interdependent components.

heating the Earth receives. The **solar constant**, defined as the irradiance of a unit of area perpendicular to the sun's rays at the mean annual distance of the Earth from the sun, is 1360 W/m^2. Since the Earth's surface area taken as a sphere ($4\pi a^2$, where a is the radius of the Earth) is 4 times its cross-section as a disc (πa^2), the average value over the surface of the Earth for intercepting solar radiation from outside the atmosphere is 1360 W/m^2 divided by 4, or 340 W/m^2. However, since about 30% of sunlight is reflected back to space without being absorbed, the mean solar heating of the Earth is about 240 W/m^2.

A change in solar irradiance of 0.1% due to the sunspot cycle results in a climate forcing of 0.24 W/m^2. Hansen and Lacis (1990) suggest that anthropogenic aerosols generate a forcing of 0.5 to 0.75 W/m^2 of cooling, but new estimates by Charlson et al. (1992) give a globally averaged forcing of 1 to 2 W/m^2 of cooling. If natural sulfate aerosols are about one-quarter of the concentration of those anthropogenic aerosols, then this would add another 0.25 to 0.5 W/m^2 of cooling. Greenhouse gas forcings resulting from the total increase of the gases from 1850 to 1989 contributed a heating of about 2.0 W/m^2. The 11-year sunspot cycle has modulated heating or cooling of the Earth by only about one-tenth of the greenhouse forcing, while aerosol cooling is about one-half of the greenhouse warming to date. Global climate models show that with a doubling of the current atmospheric CO_2 concentration, the greenhouse gas warming of the atmosphere would be 4.0 to 4.5 W/m^2. This would be equivalent to an increase in solar irradiance of 2%. A very controversial report by Seitz et al. (1989), issued by the Marshall Institute of Washington, D.C., suggested that solar irradiance can be expected to decline early in the twenty-first century and that the resulting cooling will offset greenhouse warming. Earlier we mentioned that the Gleissberg Cycle has peaked every 85 years (in 1770, 1850, and 1940). This suggests that the next peak will occur around 2025, and that there will be a slow decline of solar irradiance for about 40 years after that, followed by another increase.

It is expected from the current trends of greenhouse gas emissions that carbon dioxide or its equivalent will have doubled in concentration from the preindustrial level somewhere around 2050 to 2075. Climate models suggest an equilibrium global warming of 1.5°C to 5.5°C from a heating of 4.0 to 4.5 W/m^2 due to CO_2 doubling. The most probable temperature increase would be 3.0°C.

Climate Feedbacks

Feedbacks are reactions of the climate system to natural or human-induced change that in turn affect the pace or direction of further change. **Positive feedbacks** amplify the climate change; **negative feed-**

backs diminish it. Negative feedbacks may not reverse the change, since they result from the climate change itself, but they may reduce its magnitude.

Clouds may produce positive or negative feedbacks. Their role in the heat balance of the Earth is complex. Higher temperatures associated with the greenhouse effect may evaporate more water and produce more clouds. Low clouds have high albedos, reflect sunlight, and reduce the amount of solar heating. Low clouds also have relatively warm tops, which radiate a lot of energy out to space and thereby cool the Earth. High clouds, by contrast, reflect less sunlight, have cooler tops, radiate less energy, and thereby keep the Earth warmer.

An increase in the water vapor content of the atmosphere may also be a positive feedback by itself, since water vapor is also a greenhouse gas. Rind et al. (1991) present observational and modeling results that strongly support the idea of a positive feedback by water vapor. Their model predicts an increase in the amount of water vapor in the middle and upper troposphere, with a consequent greenhouse warming. Rind et al. used data collected by the SAGE II satellite, which has been in orbit since 1984. They compared summer and winter moisture values in regions of the middle and upper troposphere and found that as the hemispheres warm, increased convection leads to increased water vapor, in close agreement with their model. They also compared contrasting regions, such as the convective western Pacific and the largely nonconvective eastern Pacific, and got the same results.

Raval and Ramanathan (1989) approached the same problem in a different way by using satellite instrument data to measure the difference between the flux of radiation escaping to space and the flux emitted from the Earth's surface. This difference—the energy trapped in the atmosphere—is the greenhouse effect. They found that the greenhouse effect increases significantly with an increase in sea surface temperature. This finding is compelling evidence for a positive feedback between surface temperature, water vapor, and the greenhouse effect.

Other positive feedbacks include the melting of snow and sea ice that occurs with climate warming. The removal of snow from the land leaves a darker surface, which then absorbs more sunlight. Removal of sea ice leaves very dark, sunlight-absorbing seawater in place of the reflective ice. In addition, the removal of ice from the sea allows more heat to be transferred from the warm ocean to the atmosphere. Global warming may also produce a large biological positive feedback through the release of carbon dioxide and methane from vegetation, soils, and clathrates. This is a wild card in the deck, and one that is not well understood.

Kellogg (1991) discusses some of the feedbacks likely to affect the climate. And, an excellent review by Michaels and Stooksbury (1992) contains a discussion of the possibility of industrially generated sulfates providing compensatory cooling to greenhouse warming. The authors discuss many of the features I have described in the above paragraphs, including the role of water vapor; there is no doubt that clouds could play a Machiavellian role with respect to greenhouse warming. Michaels and Stooksbury state: "The fact that there are no misplacements between summer and winter in the day and night warming hypothesis further supports a relatively benign greenhouse effect."

Chaos versus Order

What we have described in the preceding paragraphs is known as a **dynamical system**. By this we mean a system whose evolution from some initial state to some later state can be described by a set of laws or rules. These rules may be expressed by mathematical equations. The system evolves slowly through a series of states, all of which may be described in a similar manner to the initial state.

The word *chaos* derives from a Greek word meaning "the complete absence of order." Many systems in nature are chaotic. There are fluctuations of temperature, precipitation, or winds that are just a part of the climate noise and are probably not predictable except in a statistical sense. Daily weather patterns exhibit a certain amount of chaos. The more long-range the weather forecast, the less likely it is to be correct. Presumably, when weather patterns are aggregated to describe a climate (the long-term averages), there is somewhat less chaos in the result.

Global Climate Models

The real world of oceans, land masses, atmosphere, snow, ice, vegetation, and clouds interacts with currents, winds, heat flow, and moisture transport resulting from solar heating. Although we can observe and measure the weather and climates of the Earth, in order to understand the dynamics of such a complex system and to be able to project the behavior of this system into the future, or even to reconstruct its past behavior, it is necessary to formulate analytical models of it.

In principle, **global climate models** or **general circulation models**—GCMs—are quite straightforward, but in reality they are complex. A model is formulated by dividing the Earth into grid boxes of approximately 5 degrees of latitude by 5 degrees of longitude (some models are 2.5 × 4 degrees); for example, a single box represents the entire state of Colorado. Each grid box is then divided into a number of horizontal layers from the ground to the top of the atmosphere, typically ten. This represents 25,920 grid points for computer calculation, just for the global atmosphere alone. The oceans are represented by grid boxes as well, each box divided into horizontal layers, including a well-mixed upper layer. The size and shape of the oceans and continents are approximately described in the model, as are the characteristics of the surface within each grid box—i.e., soil, vegetation, snow, ice, and water. Even mountain ranges are included. Cloud cover and its distribution, with height, latitude and longitude, and cloud type, are also included.

Mathematical equations describing the fundamental conservation laws of physics are solved numerically for each grid box by a computer program that calculates the transfer of mass, energy, and momentum from one box to another, and also simulates the physical processes within the boxes which represent sources and sinks of these transfers. Atmospheric GCMs can predict changes of such variables as wind velocity, temperature, moisture, and pressure. The GCMs simulate the current climate of the world reasonably well, with its horizontal, vertical, and seasonal temperature distributions, its pressure patterns and winds, and even the jet stream. The models generate the major low pressure centers that create winter storms, and the equatorial high-precipitation and subtropical low-precipitation belts. The main difficulty in the models has to do with clouds. A comparison of 14 atmospheric GCMs with clouds included revealed a sensitivity difference among them of almost threefold; this means that the way clouds are treated in the various models has a major impact on the projections of global warming.

Future Climate Projections: Evidence to Date

Global climate models have improved over the years, but they are far from perfect. GCMs forecast a continuing warming of the Earth by greenhouse gases at a rate unprecedented in human history. There are uncertainties about these projections, largely because of the deficiency of GCMs in dealing with cloud feedback processes and with their coupling of ocean and atmosphere dynamics. As modeling improves, particularly as spatial resolution becomes better and funds become available for more elaborate and longer computer time sequences, we can expect to refine our long-term projections considerably. Schneider (1990) and Kellogg (1991) review the climate debate in considerable detail.

The following sections review global changes projected under the equivalent of a doubling of the present atmospheric CO_2 ($2 \times CO_2$) and the probability of each change occurring. The evidence to date that a particular change has already begun is also given.

Extensive Stratospheric Cooling (Certain)

An increased concentration of greenhouse gases in the stratosphere will lead to more radiation loss to space and a cooling of the upper atmosphere. Reduction in stratospheric ozone will lead to reduced absorption of solar ultraviolet radiation and reduced heating. These two processes will result in a cooling of the stratosphere.

Angell (1986) shows that during the past quarter of a century the tropopause layer (9 to 16 km above the Earth's surface; see Figure 2) and the low stratosphere (16 to 20 km) have cooled and that a greater rate of temperature change with increasing altitude (the **lapse rate**) has resulted. The cooling of the lower stratosphere has been greater in the Southern Hemisphere than in the Northern Hemisphere. Combined radiosonde and satellite data indicate a significant cooling trend for the middle stratosphere (30 km) in the middle to high latitudes of the Northern Hemisphere.

Global Mean Surface Warming (Highly Probable)

All GCMs project that with a continuing increase of greenhouse gases in the atmosphere, the air temperature at the Earth's surface will become inexorably warmer. With a doubling of CO_2, or its equivalent in terms of the sum total of greenhouse gases, the long-term global mean surface temperature will increase; estimates of the amount of the increase range from 1.5°C to 4.5°C; a 3°C increase seems most likely (Figure 9). Uncertainties in the models involve possible negative feedbacks, such as clouds, and time lags in the atmosphere–ocean–land system that would reduce the rate of warming. Possible positive feedbacks include increased soil respiration in a warmer, wetter world.

Analysis of the temperature records of the world (using some 200 million temperature readings) shows a steady upward trend in the annual mean global surface air temperature; this temperature rose by about 0.6°C from 1860 to the present (see Figure 1) (Hansen and Lebedeff, 1987; Jones, 1988; Jones and Wigley, 1990; Kerr, 1990). Although the Northern Hemisphere experienced some cooling between 1945 and 1975, the warming trend has been renewed there. The Southern Hemisphere never showed a significant cooling trend. The decade from 1980 through 1990 had seven of the warmest years on record for the world as a whole. Tsonis and Eisner (1989) show that the probability of the six warmest years in the temperature record being clustered between 1980 and 1988 due to purely natural causes was between 1% and 3%. Jacoby and D'Arrigo (1989) were able to reconstruct Northern Hemisphere annual temperatures since 1671 based on tree-ring data from North America. Their study confirmed the findings of the other studies referred to above and revealed many other features in the temperature record extending back more than 300 years.

An analysis of satellite-derived sea surface temperature trends from 1982 to 1988 by Strong (1989) shows a warming trend of 0.1°C per year. Ship data shows a warming trend 0.05°C per year. (I hasten to point

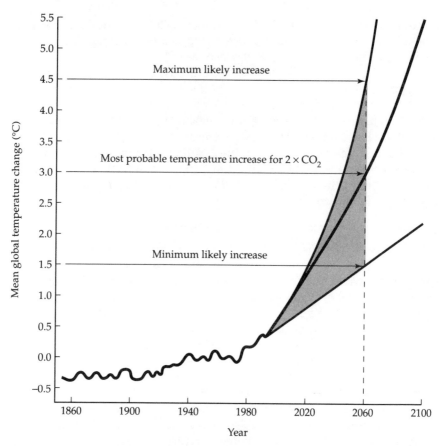

FIGURE 9 **Mean global temperature change versus year showing past temper-atures and projected most probable temperatures with increasing atmospheric carbon dioxide concentration. $2 \times CO_2$ is assumed to occur around the year 2060. Because of uncertainties with modeling and in our knowledge of the carbon cycle, maximum and minimum temperature extrapolations are also shown.**

out that short-term trends are not particularly valid; see Karl et al., 1989, for further discussion of this.) According to Trenberth (1990), the North Atlantic Ocean, and to some extent the North Pacific Ocean, have been cooling slightly during the past half-century. The greatest rates of warming have occurred over some of the continental areas at moderate to high latitudes. This is exactly what one would expect. The oceans are more resistant to temperature change because of their cir-culation patterns and their ability to store heat.

Washington and Meehl (1989) have used the National Center for

Atmospheric Research Community Climate Model (CCM—their acronym for GCM) to run an incremental experiment in which CO_2 increased gradually from year to year, as it does in the real atmosphere. This is in contrast to most GCM equilibrium models, in which a sudden increase to doubled CO_2 levels is run out until a steady-state climate condition is reached. As the CCM model's Earth grew warmer overall, the North Atlantic, including Greenland and northern Europe, cooled off significantly, particularly during the winter months. The North Pacific did the same thing, but less strongly. This cooling, which has shown up both in observations and in modeling, is caused in part by a weakening of the Icelandic low pressure cell and a shifting of the wintertime low pressure area eastward. This pattern leads to an enhanced flow of cooler air from the north, which produces a greater flow of lower-salinity water into the East Greenland Sea, increasing the stability of the upper ocean waters. The Atlantic Ocean "conveyor belt" that allows the Gulf Stream to warm northern Europe was therefore impeded. After three decades the temperature trends in the North Atlantic changed from cooling to warming (see Washington, 1990). A similar result was obtained by Stouffer et al. (1989) for the North Atlantic, and they obtained a particularly slow rate of warming for the Antarctic Ocean.

Warmer Nights (Probable)

An increase in cloudiness, humidity, and aerosols in the lower atmosphere will probably result in warmer nights due to trapping infrared heat. This effect most likely would be seen over land because the land surface has a lower heat capacity than water (it takes less heat to raise its temperature). An analysis by Karl et al. (1991) of minimum and maximum temperatures in the United States, the Soviet Union, and China during the last 40 years shows nighttime temperatures to have increased steadily by a few tenths of a degree centigrade. The *maximum* air temperatures, usually being daytime temperatures, remained unchanged or rose very little. The range of difference between the mean daily maximum and minimum temperatures decreased substantially, whereas the greenhouse effect alone should increase both maximum and minimum daily temperatures. Plantico et al. (1990) show that increased cloud cover and reduced sunshine seem to have coincided with some of this temperature change. Increased cloud cover could be a direct result of global warming.

Polar Winter Surface Warming (Very Probable)

Global climate models show the greatest rates of temperature increase occurring in the winter in polar regions. A greater fraction of open water and thinner sea ice leads to increased transfer of heat from the ocean to the air and thus a warming of the polar surface air by as much as three times the global mean warming. In summer, when ice cover is reduced or nonexistent, the additional heating due to CO_2 in the atmosphere is taken up by the ocean. The stored heat delays the appearance of sea ice in autumn and reduces its thickness the following winter. The usual seasonal cooling in the autumn is reduced by the large flux of heat from the ocean to the air. Over the continents the loss of snow cover and subsequent reduced albedo produces an early warming in the spring.

In fact, observations of land surface air temperatures show cooling in the mid to high northern and southern latitudes (Hansen and Lebedeff, 1987). Tropospheric data analyzed by Angell (1986) show slight cooling or no temperature trend during the past 30 years in north temperate, north polar, and south polar regions. The delayed warming of the North Atlantic, including Greenland and northern Europe, and the North Pacific mentioned above may slow the rate of warming for high latitudes in general. Walsh (1991) used a record of areally averaged air temperature anomalies north of 55 degrees north latitude to show that they resemble the anomalies for the Northern Hemisphere as a whole and do not indicate unusual warming in the far north.

Reduction of Sea Ice (Very Probable)

The models generally predict a reduction of sea ice with the world warming. However, new results reported by Washington and Meehl (1989), using the incremental CO_2 increase model, cast some doubt concerning sea ice reduction during the next 30 years in the North Atlantic. NOAA satellite data on sea ice extent over the last 16 years suggest no significant trend, while NASA satellite data show a declining trend. Landsat images from 1974, 1979, and 1989 show rapid distintegration of the Wordie Ice Shelf off the west coast of the Antarctic peninsula. This breakup appears to coincide with a warming trend in mean annual air temperatures in the region. (See Walsh, 1991 for further discussion of the sea ice decline and retreat of the snow-covered areas of the Northern Hemisphere during recent years.)

Global Mean Precipitation Increase (Very Probable)

The rate of water evaporation increases as the temperature rises, and this could lead to increased precipitation worldwide. GCMs show that the centers of the continents will have less precipitation than they do now, while the margins of the continents will be wetter. Manabe and Wetherald (1987) give a detailed discussion of precipitation, evaporation, and soil moisture for four regions of the world: northern Canada, the North American Great Plains, southern Europe, and Siberia. With the warming trend, northern Canada will be wetter in the winter and drier in the summer. The Great Plains should be decidedly drier in the summer, but without much change in the winter.

Observations of Northern Hemisphere land areas for the period 1950–1984 indicate a net upward trend of precipitation in the mid to higher latitudes (35°–70° N), with a declining trend in the lower latitudes (see Bradley et al., 1987 for more details). Southern Hemisphere data show a trend of increasing precipitation from 1940–1986 in both low and high latitudes, but with a lot of seasonal and regional variability.

Summer Continental Dryness and Warming (Probable)

Most GCMs indicate a marked long-term drying of soil moisture in mid-latitude continental regions during the summer. This dryness would be caused by an earlier end of snowmelt and spring rains plus an earlier onset of spring–summer drying. In Manabe and Wetherald's projections (1987), average annual precipitation increases at latitudes above 45 degrees, but because evaporation rate also increases, the net result is that soil moisture is unchanged. The long-term projection of precipitation patterns and soil moisture is among the more uncertain of the GCM outputs.

Precipitation data for the United States show above-average amounts for the two decades prior to 1984 (Bradley et al. 1987). Diaz (1983) found the interior and western portions of the contiguous U.S. to be more drought-prone than other regions. The enormous variability of precipitation patterns makes coherent comparisons with model projections difficult. Nevertheless, this is a topic of great interest to anyone concerned with ecosystems.

Rise in Global Mean Sea Level (Probable)

Sea levels are expected to rise with a warming of the world climate because of thermal expansion of seawater and increased rates of melting of continental glaciers. However, there is evidence that the warming trend at high latitudes will increase the evaporation of seawater and cause more snowfall and ice accumulation over the glaciated regions, thereby offsetting increases in surface melting. Wigley and Raper (1987) estimate that with a global warming of 0.5°C, sea level will rise 2 to 5 cm due to thermal expansion and an additional 2 to 7 cm due to glacial melting per century, based on a 10 to 15 cm per century rate of sea level rise. If the West Antarctic ice sheet melts entirely, total sea level rise is estimated at 5 m, and if all the world's ice melts, at 80 m. Such events are not likely to occur for many centuries, if ever.

Global sea levels are rising at about 24 mm per decade, according to Peltier and Tushingham (1989) and Zwally (1989). This is about double earlier determinations of the rate of rise. Experts are now estimating that this rate will increase, so that sea levels will rise about 33 cm by the year 2050, and perhaps 1.5 to 2.5 m by 2100. On the average around the world, a 33-cm rise would push shorelines inland by 30 m. Clearly, projections of future sea levels are highly uncertain.

Intensification of Coastal Upwelling (Possible)

The coastal ocean off the west coasts of the United States, the Iberian Peninsula, southwestern Africa, and Peru is a wind-driven coastal upwelling system in which colder waters from the ocean's depths are brought close to the surface. The vigorous alongshore wind that drives this coastal upwelling is maintained by a strong atmospheric pressure gradient between a thermal low-pressure cell over the warm land mass and higher pressure over the cooler ocean. Areas inland of these foggy coastal regions are under the influence of strong atmospheric subsidence (sinking of air) and are characterized during upwelling seasons by dry Mediterranean-type or desert climates and clear atmospheric conditions. With greenhouse warming, we expect an inhibition of nighttime cooling and an enhancement of daytime heating, which should lead to intensification of the continental thermal low-pressure cells adjacent to upwelling regions. This would result in increased onshore-offshore atmospheric pressure gradients, intensified alongshore winds, and accelerated coastal upwelling.

Bakun (1990) gives evidence of increased coastal upwelling, as deduced from increased wind stress, off the west coasts of the United States, the Iberian Peninsula, southwestern Africa, and South America. Intensified upwelling would enhance primary productivity in these waters and could lead to increased fish production.

The Rest of the Story

The predicted unprecedented rate of global warming and dramatic changes in precipitation patterns will strongly affect agriculture and ecosystems. A warming of 3°C over the next 60 to 90 years suggests a rate of displacement of the global mean annual isotherm of 0.5 to 0.33°C per decade, although the displacement will vary considerably in different parts of the world. A 3°C temperature change in the central United States will move the mean annual isotherm by 250 km northward and up mountain ranges by 500 m. This represents a movement rate of 28 to 42 km per decade.

Global climate modeling shows that with global warming, on an annual average there will be a drying trend at middle latitudes. This dryness would have considerable impact on agricultural regions, a subject discussed in more detail in Chapter 7. A warmer world is likely to have a good many more storms than the present world, because more energy will be available to generate increased turbulence in the atmosphere. More frequent tornados may occur, there may be an increase in the frequency of hurricanes and typhoons, and wet regions may get more frequent deluges. More dryness during all or part of the year may increase the frequency of fires in many parts of the world.

All these possible changes and many more are discussed in the remaining chapters. First we consider climates of the past, their changes, and the forces that produced those changes. We consider how the ecosystems of the past responded to paleoclimate changes in order to better understand the future. Then we take up the likely effects of projected climate changes on both natural and agricultural ecosystems. Not all of these effects are harmful. There are many possible favorable changes, such as increasing CO_2 levels improving photosynthetic productivity and water use efficiency by plants, or increasing temperatures favoring some organisms, or increased precipitation improving agricultural ouputs. However, we must remind ourselves that inexorable forces may drive the climate system too quickly into conditions that are too severe for most ecological adaptation or adjustment to take place. Then there is always the question of the human condition: whether our food supply will be adequate; whether resources will be renewable or sustainable; whether the climate will provide comfort or unmanageable discomfort; and, finally, whether the political, social, and economic systems devised by human mentality will allow for sufficient mitigation.

C H A P T E R   T W O

Past Climates

THERE IS MUCH to be learned by studying climates throughout the course of the Earth's history. In fact, if we are to understand how and why climates change now and in the future, we must be able to understand the climates of the past. Fortunately, today we are able to simulate present and past climates by means of the global circulation models (GCMs) discussed in Chapter 1. Kutzbach (1985) has reviewed the development of modeling and, in particular, its application to the interpretation of paleoclimates.

During the last part of the Mesozoic era, 100 to 65 million years ago, the world was more than 20°C warmer than it is today (and it was even warmer during earlier times). It was a world without ice caps. About 55 million years ago the global climate began a long, slow, cooling trend. The present "glacial" age is thought to be at least the third time that the Earth has experienced widespread continental glaciation; the earliest was about 600 million years ago and the second about 300 million years ago. During these earlier times the Earth's land masses were joined in a single supercontinent, called Pangaea. Then continents moved apart as the result of sea-floor spreading and continental drift. As land masses separated and drifted apart, over periods of tens of millions of years, ocean circulation patterns changed, and this had great effect on the climate. How do we determine what the climate was like, both at that time and more recently? How do we infer the temperature and precipitation patterns of long ago? How do we determine the ages of fossils, rocks, or sediments?

Proxy Thermometry and Age

Since the invention of the thermometer, the record of air temperature measured near the Earth's surface has become increasingly exact, and with satellite radiometry, many gaps, such as areas over the oceans,

are being filled. To infer the world's temperature prior to the instru-
mental record one must rely on a variety of proxy methods. The annual
growth of tree rings, annually layered lake sediments (known as
varves), and isotopes in ice cores may be correlated with modern in-
strumental temperature measurements and then used to extend the
record as far as 10,000 years into the past. These records can give an
indication of annual temperatures. Beyond that, and even within that
time frame, studies of pollen concentration in lake deposits; fossil organisms in soils, sediments, or rocks; and oxygen isotopes in ocean
sediments may be used to extend the temperature record back several
hundred thousand or even millions of years. These records can give an
indication of temperatures over decades, centuries, or longer periods.
Associations of certain flora and fauna in the fossil record are also a
useful means of reconstructing paleoclimates. Many organisms, from
microbes to large mammals and from plankton to forest trees, have
well-defined climate requirements, so when they are found in fossil
remains, the broad outlines of the climate they lived in can be inferred.

Isotopic Temperatures

The story of the discovery of isotopic temperature determination is a
fascinating one and is worth repeating here in brief. Harold Urey, who
won the Nobel Prize for his discovery of heavy hydrogen (deuterium),
was giving a lecture at the Technische Hochschule in Zurich, Switzerland in 1946. He pointed out during the lecture that although isotopes
of an element behave the same chemically, they have different masses
and therefore act differently in physical processes. Water molecules
(H_2O) containing the three isotopes of oxygen (of atomic masses 16, 17,
and 18; see Figure 1) evaporate at three different rates. Water molecules
with the lighter isotopes will vaporize faster than those with the heavier
isotopes. Therefore, over time a body of water becomes slightly enriched
in the rarer, heavier isotopes. The oceans, long subjected to this evaporative separation, should be slightly richer in heavier water than would
fresh rainwater, for example.

Professor Paul Niggli, a distinguished Swiss crystallographer,
pointed out to Urey that the skeletons of organisms living in seawater
should contain the same isotope ratio as the water in which they are
living. If this is so, then an analysis of carbonate deposits, such as
limestone, coral, or skeletons in sediments, should reveal whether the
organisms originated in marine or fresh water. Urey later calculated
that the isotope ratio would depend on the temperature of the water—
that warmer water would evaporate the lighter isotopes faster, leaving
the more dense water behind. As Urey said later, "I suddenly found
myself with a geologic thermometer in my hands."

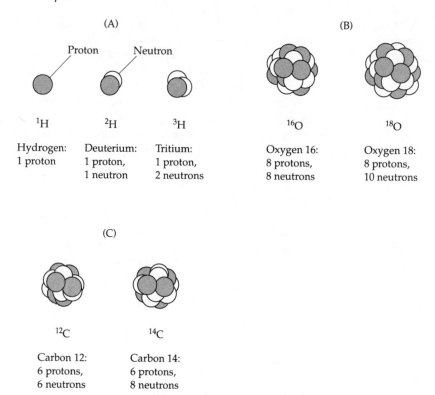

FIGURE 1   A chemical element is characterized by the specific number of protons in its nucleus; all atoms of oxygen, for example, contain 8 protons; any other number of protons would define a different element. The number of neutrons, however, can vary from atom to atom of the same element. These variants are known as isotopes, and are a useful tool in scientific research. (A) Some isotopes of hydrogen. (B) The oxygen in water molecules usually contains eight protons and eight neutrons (oxygen 16); Harold Urey and Paul Niggli made use of the oxygen 18 isotope in their calculations to measure changes in temperature over time. (C) Carbon 12, the common form, has six protons and six neutrons. Carbon 14 has eight neutrons and is radioactive; it is a useful tool in dating organic remains.

But Urey was still far from developing a useful measuring tool. The isotope ratios had to be measured with enormous precision, which meant improvements in the mass spectrometers used at that time. Next a temperature calibration was needed. Urey and his associates grew marine mollusks in the laboratory at fixed temperatures and measured their oxygen isotope ratios. Little by little they arrived at a temperature calibration for marine invertebrates, including ancient belemnite fossils.

Finally they fine-tuned their thermometer by testing the skeletons of foraminifera—tiny single-celled marine animals. Most species of foraminifera live on the deep sea bottom, but some of them float in the upper sunlit strata of the oceans, where they feed on microscopic plants. When foraminiferans die their skeletons of calcium carbonate drift down to the ocean bottom and form sediments. The skeletons build up layer upon layer, and although only a couple of centimeters accumulate every thousand years, they eventually build up to thicknesses of hundreds of meters or more. When these long-protected sediments are recovered in drill cores from deep ocean basins, their oxygen isotope ratios can be measured.

The oxygen isotope composition of a sample is usually expressed as the departure of its oxygen 18/oxygen 16 ratio from an arbitrary standard. The values are expressed per thousand units and written as follows:

$$\delta^{18}O = \frac{(^{18}O/^{16}O)\text{sample} - (^{18}O/^{16}O)\text{standard}}{(^{18}O/^{16}O)\text{ standard}}$$

Negative values mean the sample is depleted of oxygen 18 and positive values mean it is enriched or isotopically heavier. The sample value is then compared to a standard value. Two standards used are (1) a standard mean ocean water (SMOW), or (2) Peedee belemnites (PDB). (Peedee is the site location for the fossil organisms used to obtain the standard.) Usually the best practical temperature resolution is about 0.5°C. A $\delta^{18}O$ value of -10% (parts per thousand) indicates a sample with an $^{18}O/^{16}O$ ratio of 1% less than the standard. The lowest $\delta^{18}O$ ratio ever reported is 5.7% depletion, for a sample of snow from a remote part of Antarctica, meaning that more ^{18}O was left behind in the seawater during evaporation—and that the water was very cold.

Isotopic Dating

To make inferences about past climates, it is not sufficient to know only the temperature at which the fossil organism lived; it is also necessary to determine when they lived. Fortunately, an understanding of nuclear physics provided just the clocks for rocks and organic remains. Elements or isotopes that have unstable nuclei undergo spontaneous disintegration, or radioactive decay, and in the process change into other elements or isotopes. The rate of radioactive decay is constant for a particular isotope and is unaffected by changes in chemical or physical conditions. The time it takes for a radioactive isotope to decay until half the initial amount remains is called its **half-life**. For radioactive carbon (radiocarbon, or carbon 14; Figure 1C), the half-life is 5730 years.

Carbon 14 is produced naturally in the atmosphere by the bombardment of nitrogen by cosmic ray neutrons. Plant remains in old sediments, soils, bogs, timbers, middens, or other places are used for radiocarbon dating. The principle assumes that all plants assimilate carbon dioxide from the atmosphere and that carbon 14 is assimilated at the same rate as carbon 13 and carbon 12. However, only carbon 14 is radioactive and therefore unstable. During photosynthesis, an isotopic fractionation occurs because carbon 12 is more readily assimilated than either carbon 13 or 14. When a plant dies, photosynthesis stops and no new carbon is assimilated. At this point the plant has approximately the same proportion of carbon 14 as the atmosphere, but as time passes the carbon 14 decays back to nitrogen with the emission of a beta particle (an electron); carbon 12 does not decay, however, so we can determine the amount of carbon 14 that would have been present when the plant died by calculating it as a fraction of carbon 12. The amount of carbon 14 remaining is assessed by using a Geiger counter to count the beta rays emitted from the sample. Once the amount of carbon 14 remaining in an ancient sample is measured one can determine its age by the time it took to decay to that amount.

Carbon 14 can only be used for dating back about 75,000 years. Beyond that, the amount of radioactivity remaining is too little to be accurately counted. See Grootes (1978) and Stuiver et al. (1978) for more details.

Sun–Earth Geometry and Climate

We know from everyday experience that the sun has a lot to do with the climate on Earth. Solar radiation is the energizer of the Earth's climate. We know it warms the equatorial regions more than it does the polar regions of the world, and that the tilt of the Earth from the plane of the ecliptic causes winter in one hemisphere while the other hemisphere experiences summer.

The Earth's Motion

Three characteristics of the Earth's motion in orbit around the sun have been thought to influence the amount of radiation incident on the Earth and its distribution with latitude. The first is eccentricity, the second is obliquity, and the third is precession.

Johannes Kepler (1571–1630), the distinguished German mathematical astronomer, discovered that the Earth moves in an elliptical orbit around the sun, which brings the Earth slightly closer to the sun at one time of year and further away at another time. Since the amount of solar radiation striking the Earth varies inversely with the square of the distance from the sun, the Earth receives more radiation at perihelion (its closest approach to the sun) and less radiation at aphelion (farthest distance). It was almost 200 years after Kepler that the French astronomer Urbain Leverrier discovered that the shape of the orbit gradually changes over time from more elliptical to more circular. This change has a periodicity of about 100,000 years, and is the result of the pull on the Earth by the other planets. The measure of this ellipticity is called **eccentricity** (Figure 2A).

The axis on which the Earth spins is not at right angles to the plane in which it revolves around the sun (the *ecliptic*), but is tilted at 23.5°—the angle between the plane of the equator and the ecliptic. This tilt angle is referred to as the Earth's **obliquity** (Figure 2B). The obliquity varies from 22.1° to 24.5° with a periodicity of 41,000 years. This change affects the latitudinal distribution of the solar radiation striking the Earth.

Around 125 B.C. the Greek astronomer Hipparchus discovered that the Earth's North Pole does not always point in the same direction among the stars. The pole wanders, or **precesses**, slowly, with a periodicity of 22,000 years. The motion of the Earth is like that of a spinning top with its axis inclined to the vertical and slowly describing a cone in space (Figure 2C). The cone has an angular radius equal to the obliquity, or 23.5° at the present time. Today the North Pole points nearly at the "pole star" in the constellation Ursa Minor, but 4000 years ago it pointed near a star in Draco, and 5000 years in the future it will point near a star in Cepheus. The constellations seen 5000 years ago from a given position on Earth were very different than they are today. For example, in 3000 B.C. the Southern Cross constellation, which today cannot be seen from the United States or Canada, could have been seen from north of Quebec. The cause of precession is the pull of the sun and moon on the protruding material, or bulge, at the Earth's equator. Precession affects the position of the equinoxes and solstices as they shift slowly around the Earth's elliptical orbit, completing one full cycle every 22,000 years. Today the winter solstice occurs near the perihelion (closest to the sun). Eleven thousand years ago, the winter solstice occurred near the opposite end of the orbit.

The Ice Ages

Many people have wondered what caused the ice ages. In 1875 the Scotsman James Croll published an important book that reviewed many of the earlier ideas concerning the orbital motion of the Earth and climate. Although Mairan in 1765 and Lagrange in 1782 appreciated the importance of eccentricity to the Earth's climate, the imprecision of data concerning the planetary parameters and poor information on geological

(A) ECCENTRICITY

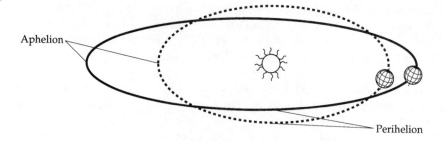

Aphelion

Perihelion

(B) OBLIQUITY

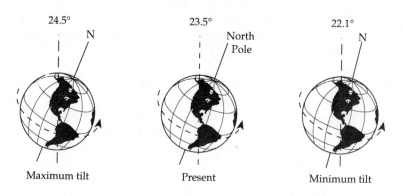

24.5°
N

Maximum tilt

23.5°
North
Pole

Present

22.1°
N

Minimum tilt

(C) PRECESSION

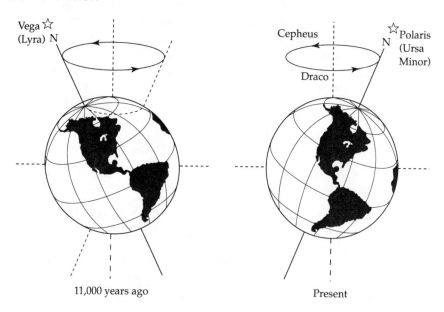

Vega ☆
(Lyra) N

11,000 years ago

Cepheus

Draco

☆ Polaris
N (Ursa
Minor)

Present

events made it difficult to arrive at correct conclusions in the eighteenth century. Croll speculated that if precession caused the Earth's axis to tilt directly away from the sun at the same time the orbit was highly eccentric, thus carrying the Earth further from the sun, winter cold would be at its extreme. If colder winters meant more snowfall, this would reflect more sunlight and make things even colder, gradually initiating an ice age.

Croll put together the 22,000-year precession cycle and the 100,000-year eccentricity cycle and obtained a pattern thought to resemble that of the ice ages. Although he was right in some respects, when more information concerning the occurrence of ice ages was found, his calculations would not fit and his idea was discarded. The Yugoslav mathematician Milutin Milankovitch revived Croll's hypothesis many years after Croll's death in 1890. Milankovitch added the 41,000-year obliquity cycle to Croll's two cycles, and after years of calculations he came to the conclusion that only the precession and obliquity cycles were important in changing the climate. Milankovitch published his ideas in 1930. He had no proof for his model and left it up to geologists to find the evidence.

After Urey had established the oxygen isotope method of inferring temperatures from deep-sea deposits of foraminifera, efforts were made to obtain cores of ocean sediments. Measurements from these cores could be interpreted in terms of ice volume and dates. At first there was a lot of disagreement and controversy over interpretations, but by the 1970s the picture began to clarify. At the heart of the solution to the ice age problem was a special sediment core taken from the Indian Ocean in 1971, with layers dating back 450,000 years. The global ice volume was inferred from the $\delta^{18}O$ temperature measurements in the sediments of this core. Hays et al. (1976) modeled the orbital cycles and compared them with the temperature patterns inferred from cores. They found correlations not only with the very strong eccentricity cycle, 100,000 years, but also with the 41,000-year obliquity *and* the 22,000-year precession cycles. This 22,000-year cycle is actually a quasi-periodic precession of about 19,000 and 23,000 years. This confirmed, for the

◄ FIGURE 2   (A) Eccentricity. The Earth's orbit varies from extremely elliptical to nearly circular over a period of some 100,000 years. (B) Obliquity. The tilt of the Earth's axis varies from 22.1° to 24.5° over about 41,000 years. Presently this tilt is 23.5°. (C) Precession of the North Pole. As the Earth spins on its axis, it slowly describes a cone in space. Over thousands of years the "pole star," or position in space to which the pole points, will change; it will return to its present pole star in about 22,000 years.

(A)

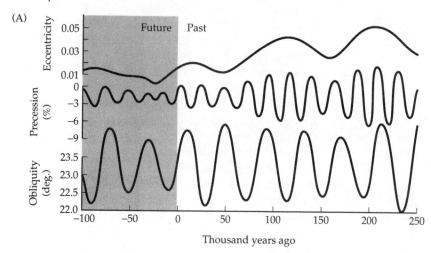

Thousand years ago

(B)

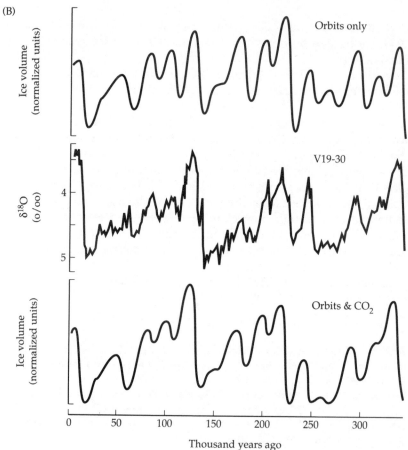

Thousand years ago

first time, the Croll–Milankovitch solar hypothesis for the origin of the ice ages (Figure 3).

Imbrie and Imbrie (1980) made improvements in the orbital model but also described some of its shortcomings when applying it over a full 500,000-year time span. The model's simulation of the isotope record of the ice volume over the last 250,000 years is quite good, and is fairly good out to 350,000 years, but it is decidedly less accurate beyond that. The problem is related to the generation of the 100,000-year eccentricity cycle.

Carbon Dioxide, Methane, and Ice Age Climates

The close association between greenhouse gases and glacial–interglacial climate changes was shown by an analysis of air trapped in Greenland (Camp Century and Dye 3) and Antarctic (Byrd, Dome, and Vostok) ice cores. Only the Vostok core gives a complete, undisturbed series over the last 150,000 years (Figure 4). Earlier, Broecker (1982) had suggested that an increase in the atmospheric CO_2 concentration of about 100 ppm at the end of the last glacial stage may have provided an amplification of the very rapid melting of the North American ice sheets through greenhouse warming. Glaciers melt much more quickly than they form, but the orbital cycle changes of solar irradiance could not account for this rate of climate change. Pisias and Shackleton (1984) showed that by adding an atmospheric CO_2 forcing to a model of orbital global climate change, they could get a strong 100,000-year cycle and also fit most substages of the ice volume record (Figure 3B, bottom trace). The new data shown in Figure 4 suggest that methane also plays a strong role.

The atmospheric CO_2 content was 190 to 200 ppm during the last glacial maximum, compared with 270 ppm in the mid-Holocene until the Industrial Revolution, and 355 ppm today. The CH_4 concentration was 350 ppb during the glacial maximum and 650 ppb after the ice melted. Today it is about 1700 ppb and increasing at 17 ppb per year. This is in contrast to the most rapid rates of increase during deglaciation, which were 0.2 to 0.3 ppb per year. Throughout the Vostok ice core record there was a remarkable correlation between atmospheric CO_2, CH_4, and temperature change. There are, however, differences observed over time between the CO_2 and CH_4 changes that imply different

◄ FIGURE 3 Relationship of orbital cycles to climate. (A) Variations in orbital geometry as a function of time, both past and future. (B) An ice-sheet model output using orbital changes plus changes in the CO_2 concentration. The ice volume record as inferred from deep-sea core V19-30 sediment record is shown in between.

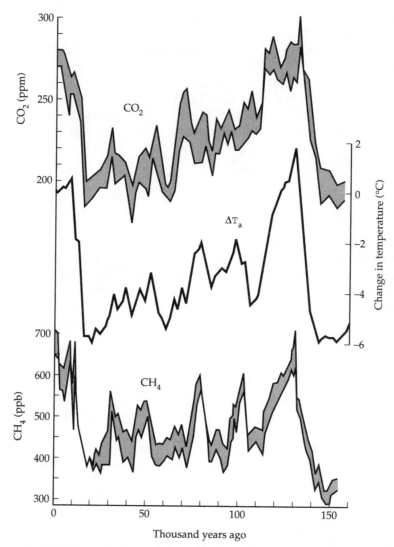

FIGURE 4   Variations in the amounts of carbon dioxide and methane, and the change in atmospheric temperature (ΔT_a) over Antarctica derived from measurements along a 2083-meter ice core from Vostok Station for the last 150,000 years. This record was reported by Chappellaz et al. (1990) and put into the form shown here by Lorius et al. (1990).

sources and sinks for the two gases, such as ocean productivity for CO_2 and wetlands emissions for CH_4. All the orbital cycles are present in the ice variation over time and in the temperature record. They also appear in the variation with time of the CO_2 and CH_4 concentrations.

The changes in the abundance of these gases are in some way being driven orbitally, but are not completely understood.

If the oceans were a lifeless puddle of water beneath the atmosphere, carbon dioxide would distribute itself between water and atmosphere solely on the basis of the chemical solubility of the gas. If this were the case, the ocean would give up much of its carbon dioxide to the atmosphere, tripling its concentration there. However, microscopic plants and animals in the surface waters of the oceans take up carbon dioxide and build tissues and skeletons from it. When they die their carbonate remains sink into the deep sea. Once there, much of the carbonate dissolves, and it cannot return to the surface waters because of the large temperature and density difference between the cold bottom water and the warm surface water. This biological pump acts to move carbon from the Earth's surface and atmosphere to the deep sea. The speed of the biological pump is recorded in the carbon isotopes of the carbonate skeletons in the sediments. Organisms prefer the lighter carbon 12 isotope over the heavier carbon 13 isotope during their assimilation of carbon dioxide. Therefore, the biological pump leaves the heavier carbon in the surface water and carries the lighter carbon to the deep water.

Chappellaz et al. (1990) speculate that monsoon precipitation patterns may have had a primary role in controlling paleomethane concentration through their effects on low-latitude wetland areas. Increased rainfall would increase the extent of wetlands, and CH_4 emissions from wetlands increase with warmer temperatures (an example of positive feedback; Chapter 1). Nisbet (1990) speculates that increasing solar insolation during deglaciation destabilizes the methane stored in northern gas fields and gas clathrates during glaciation. The primary sink for methane in the atmosphere is in reaction with OH radicals to form water and CH_3.

The mean global warming during the last deglaciation is estimated at 4.5°C. This temperature change may have resulted from increased solar insolation caused by orbital cycles, plus a climate forcing by changes in the CO_2 and CH_4 concentrations. Of the 4.5°C temperature increase, 2.3°C may have come from a climate forcing by these two greenhouse gases.

Paludification

Klinger (1991) has developed a hypothesis that ecological succession proceeds from grassland to woodland to peatland during interglacial periods. This conversion of forests to peatlands is called **paludification** and depends primarily on the growth of mosses. Paludification results in increased soil acidity, evapotranspiration, albedo, and carbon storage. Such changes would have a direct effect on climate. During the early stages of peatland formation extensive forest death would result in the decay (oxidation) of large quantities of wood, which would release CO_2. Peat formation during the same period would partially compensate by storing carbon. The long-term effect of peat accumulation would be a reduction in the atmospheric CO_2 concentration. During the early stages of paludification, emissions of CH_4 would increase as anaerobic soils allow for bacterial decomposition. However, during late stages, as *Sphagnum* mosses become dominant, the acidity of the soils would increase, suppressing bacterial activity and thus reducing CH_4 release. This reduction of atmospheric CO_2 and CH_4 during widespread peatland formation would then reduce atmospheric temperature through decreased greenhouse warming and promote another ice age.

During glaciation periods, peatlands are mechanically destroyed by the advancing glaciers. The dying organic matter that is oxidized would release CO_2 and CH_4 to the atmosphere and produce a positive feedback on greenhouse warming. Rapid and extensive deglaciation would then return the land to the process of succession from grassland to woodland to peatland once again. We see here another example of the Gaia hypothesis of Lovelock (1979).

When will the next ice age take place? If past ice ages indeed follow the orbital forcing cycles (including the CO_2 and CH_4 concentration changes), then they can be used to anticipate the next glaciation, provided that human-induced changes do not override them. The present interglacial should prevail for another 2000 years, perhaps less, and should give way to a slow cooling, bringing moderate cold 5000 years from now and major ice ages about 23,000 and 60,000 years in the future.

Ancient Climates

The Last Million Years

The Earth's climate during the last million years is best characterized as an alternation of glacial and interglacial episodes. In the Northern Hemisphere the ice sheets formed, advanced, and retreated. During the last million years there has been a series of relatively rapid, strong fluctuations in glacial ice volume, with 10 major and 40 minor ice ages. The record of these fluctuations can be found in diverse and widely distributed records, including the chemical composition of Pacific sediments, skeletal plankton remains in Caribbean sediments, and soil profiles in Europe. The ice volume as well as temperature can be estimated from the oxygen isotope record of ocean sediments (Figure 5). Only rarely during the last 850,000 years did the global temperature approach that of the present warm interglacial period.

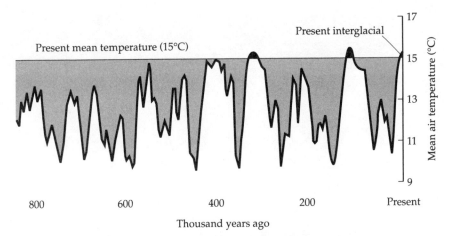

FIGURE 5 **Temperature of the Earth for the last 850,000 years as inferred from ice volume derived by oxygen isotope measurements from ice cores.**

The Last 150,000 Years

The last time the world's climate resembled the warm, relatively ice-free condition of the Holocene was during the Emian interglacial, about 130,000 years ago (Figure 6). The warmest part of this period lasted for only about 10,000 years. This reminds us that the warm interglacial Holocene time in which we are living today is well beyond its halfway point. The Emian interglacial coincided with the summer solar insolation peak for 65 degrees north latitude, which occurred 128,000 years ago. At times since then, it has been as much as 6°C colder than the present mean temperature.

One of the most detailed sets of proxy climate records for the last 150,000 years is that of African and Indian monsoonal circulations. The term **monsoon** refers to large-scale, seasonal wind and precipitation patterns, such as those observed over tropical Africa and India. The annual cycle of solar radiation heats the African and Asian continents during the Northern Hemisphere summer season and cools them during the winter. The heat storage capacity of the ocean is much greater than that of the land, so the land becomes much warmer than the ocean in summer and much cooler in winter. The vigorous summer heating of the land produces ascending air over the land and an inflow of air from the ocean to the land. This inflow of warm, moist air produces a low-pressure cell over northern Africa and southern Asia and results in heavy summer precipitation in these regions—the monsoon rains.

In the tropics the precipitation patterns and wind directions are

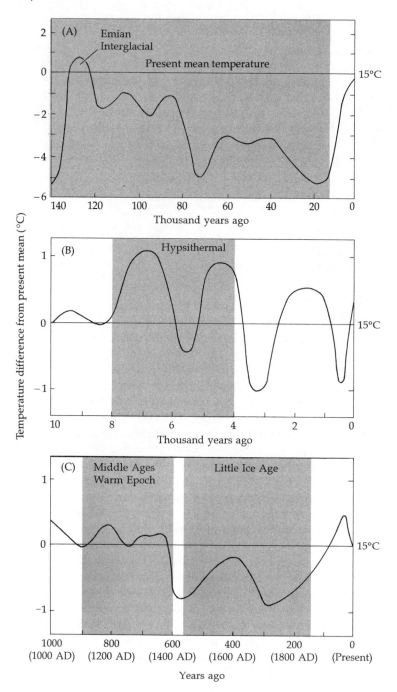

◄ FIGURE 6 (A) The mean global temperature of the ocean during the last 140,000 years, determined from isotope data and flora and fauna records in ocean sediments. The Pleistocene Epoch (gray shading) ended about 12,500 years ago, to be followed by the Holocene, or Recent. (B) Air temperatures for eastern Europe during the past 10,000 years, determined from tree ring data. (C) Mid-latitude air temperatures for the Northern Hemisphere for the past 1000 years.

recorded in many types of geophysical and biological processes. In the Atlantic Ocean, the variable occurrence of wind-blown freshwater diatoms in deep-sea sediments relates to changes in African monsoons. High densities of diatoms imply intervals of aridity and low lake levels; low diatom densities indicate more moisture and higher lake levels. Layers of rich organic sediments, called sapropels, in the eastern Mediterranean Sea indicate maximum tropical African monsoon runoff via the Nile River. Pollen transported from the African continent by southwest winds is deposited in the Gulf of Aden sediments. The pollen record can be interpreted to indicate a combination of changes in precipitation and vegetation in northeast Africa and in the strength of southwesterly Indian monsoon winds. More pollen indicates wetter conditions and a stronger monsoon. The summer monsoon winds cause ocean currents to transport high-salinity water from the western Indian Ocean to the southern Bay of Bengal. Off Arabia, southwesterly winds cause coastal upwelling with low sea-surface temperatures and high nutrient flux. This is recorded in the plankton species deposited in the sediments.

Paleoclimate data and GCM modeling show weaker monsoons during the last glacial maximum (18,000 years ago) and stronger monsoons during the early Holocene (9,000 to 10,000 years ago). Prell and Kutzbach (1987) modeled the climate of the last 150,000 years and compared patterns of average Northern Hemisphere summer radiation with many paleoclimate records for this time period. All of the paleoclimate indicators mentioned above, along with an oxygen isotope record for temperatures and inferred ice amounts, plus the estimated solar radiation averaged for the Northern Hemisphere summer, show two major patterns. First, four strong monsoon-related events in southern Asia and equatorial North Africa coincide approximately with the four major maxima of the summer solar radiation at 10, 82, 104, and 126 kya (thousands of years ago). Second, the paleoclimate records and model simulations indicate that southern Asia was drier, with weaker monsoons, when the radiation maxima were not as strong—between 75 and

15 kya—and when the extent of the ice sheets was greatest. Prell and Kutzbach demonstrated that climate modeling combined with proxy paleoclimate data can give us a detailed record of the past climate with some confidence.

The Last 18,000 Years

During the last 18,000 years, the climate changed from full glacial to full interglacial. It is for this time period that the best proxy records of past climates and chronology exist. Massive ice sheets covered parts of North America and Europe 18 kya (Figure 7A); sea ice extended at least 10 degrees of latitude closer to the equator than it does now; ocean temperatures were cooler; boreal grasses, sedges, and conifers covered latitudes south of the glaciers; and the tropics were drier than they are today. Rapid melting of the glaciers took place between 15 and 12 kya (Figure 7B), as solar insolation increased and conditions warmed. By 9 kya the ice sheets had disappeared except in northeastern North America, and sea level and sea-surface temperature approached today's values. It was warmer and drier than at the present in the northern mid-latitudes from 9 to 6 kya; it later became cooler and more moist. During all this time drastic changes in vegetation distributions took place, large numbers of Pleistocene fauna (including the mammoth and the mastodon) became extinct, agriculture was developed, and humans migrated into North America.

One of the great debates in modern anthropology is concerned with the dates for the migration of people into North America. At the time of maximum ice extent, sea level was 100 meters lower than it is today and a vast corridor, Beringia, connected Asia and northwestern North America. Humans were in northeastern Asia around 25 to 30 kya. Most evidence in the New World for human settlement dates to about 12 kya or a little earlier. One must wonder why people arrived near the northwestern edge of North America as early as 30 kya but did not cross the land bridge to occupy the vast continent existing to the east. Perhaps these people were confronted by the Cordilleran ice sheet—a monolithic, mountainous block of ice that would have prevented their passage from unglaciated Alaska into the northwestern United States. On the other hand, there is plausible evidence for human occupation at the Meadowcroft Rock Shelter site in Pennsylvania as early as 19,600 years ago. If the Cordilleran and Laurentide ice sheets coalesced during the peak of the glaciation, when did a corridor between them open up a path for human migration?

By means of GCM climate modeling, we now have answers to some previously unanswered questions: Why were lake levels high in the American Southwest during the glacial maximum and low in the North-

(A) 18 kya

(B) 12 kya

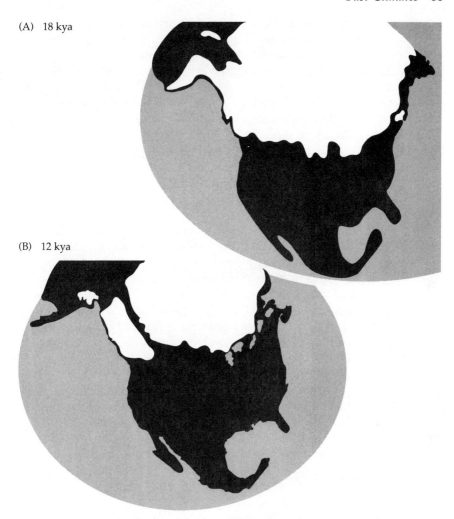

FIGURE 7 (A) Extent of the North American ice sheet 18 thousand years ago. (B) The North American ice sheet at 12 kya.

west? Why were lakes in the Sahara filling with water during the early Holocene (12 to 6 kya) when all the American desert lakes were dry? Why did mid-latitudes in North America become warmest in the period 6 kya known as the **hypsithermal**, when the summer solar radiation maximum was 9 to 11 kya?

The way to reconstruct an ice age climate is by using GCMs and then carefully comparing the model outputs with observed data. On a global grid, GCMs are used to model numerically a time-averaged, three-dimensional atmosphere in equilibrium with the following boundary conditions: the extent and elevation of the ice sheets; the sea-surface temperature; the shape, extent, and height of the continents; the surface albedo; and the concentration of atmospheric carbon dioxide. Early modeling results were reported by Gates (1976) and by members of the Climate Long-Range Investigation Mapping and Prediction Project (CLIMAP) in 1976. More recently, the Cooperative Holocene Mapping Project (COHMAP) members (1988) had as their goal "an improved understanding of the physics of the climate system, particularly the response of tropical monsoons and mid-latitude climates to orbitally induced changes in solar radiation and to changing glacial-age boundary conditions, such as ice sheet size."

Any model must be able to simulate present-day conditions before it can be used to suggest what ancient climates were like. The GCM used for the COHMAP studies reproduced the position and westerly flow of the jet stream across North America and Europe. It reproduced the Atlantic and Pacific subtropical high-pressure cells that dominate summer circulation, and the Aleutian and Icelandic lows over the northern oceans in the winter. The seasonal and latitudinal distribution of solar radiation 18 kya was similar to that of today. Any differences between climates then and now must therefore be the result of altered surface boundary conditions. These differences are illustrated in Figure 8. The glaciers and sea ice were at their maximum extents 18 kya. The sea-surface temperature was 10°C colder than it is today in the North Atlantic and North Pacific, but the global mean annual sea-surface temperature was 4°C colder. In the model, the ice sheet was kept at its maximum until 15 kya and then allowed to decrease until after 9 kya, with permanent ice remaining only in Greenland. The atmospheric CO_2 concentration was taken as 200 ppm at 18 kya, but was given the present-day value after that.

Proxy climate data were obtained in several ways. Lake levels were reconstructed using a variety of geological (ancient shorelines), sedimentary, and paleoecological data. Closed-basin lakes are good indicators of the balance between precipitation and evaporation, which, in turn, are coupled to the winds. Pollen accumulates in lake sediments and can be used to reveal the plant genera growing nearby at times past. The distributions of sedges, grasses, shrubs, and forest trees can be located from pollen deposits. From these can be inferred whether conditions were wetter, drier, colder, or warmer when the particular plants were growing. Fossils in sediments are also useful indicators of past climates.

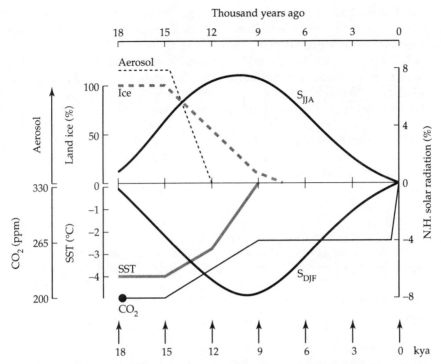

FIGURE 8 **Boundary conditions for the COHMAP simulation of the last 18 thousand years. External forcing by solar radiation is shown for the Northern Hemisphere in June through August (S_JJA) and December through February (S_DJF) as percentage difference from the radiation at present. Internal boundary conditions include land ice as a percentage of the 18 kya ice volume, global mean annual sea-surface temperatures (SST) as departures from present, excess glacial-age aerosol with arbitrary scale, and atmospheric CO_2 concentration in ppmv. Arrows mark the times of the seven simulation runs. The simulation for 18 kya included a lowered CO_2 concentration (200 ppm), while the others had 330 ppm.**

18,000 years ago Modeling showed that at 18 kya the massive, 3-km-thick ice sheet in North America caused the winter jet stream to divide around it with a strong westerly flow aloft north and south of the ice. The simulated temperatures over the land were much lower than at present, especially over the ice sheets. Strong anticyclonic (clockwise) circulation over the Laurentian ice sheet brought cold air to the North Atlantic south of the ice. This resulted in dry conditions in the American Northwest, where lake levels were very low. Farther south, westerly flow brought moist Pacific air over the land and generated the high lake levels found in the proxy record. In North America, spruce forests dominated the Midwest just south of the ice, subalpine parkland prevailed in the Pacific Northwest, and tundra in Alaska. Northern pines grew farther south than today in the Midwest, and oak grew in the Southeast. Spruce and oak forests were absent from Europe, where permafrost was widespread and conditions were cold and dry.

The GCM-simulated climate of 18 kya, with very cold air over the glaciers, had a strengthened north–south temperature gradient over Eurasia, and a displaced polar front and westerly flow to the southward over northwestern Africa, the Mediterranean, and southern Asia. The circumantarctic sea ice extended about 7 degrees of latitude closer to the equator than it does today, and this produced a steeper temperature gradient that intensified the southern westerlies across the Indian and South Pacific oceans and weakened the monsoons. The tropics were cooler than they are today, but with similar rainfall. Eastern Africa was wetter and southern Asia was drier. The paleoclimate data show lakes to have been dry in the Sahara. Weaker monsoonal activity was indicated by the foraminifera and pollen abundances in the western Arabian Sea.

12,000 years ago   By 12 kya the climate was warming as the summer solar radiation had increased by 7%. The ice was retreating, and the winter jet stream no longer divided around it. The anticyclone over the ice had weakened enough for westerlies to displace easterlies in the western United States. Lake levels were beginning to drop in the western United States, probably because of high insolation. Europe was warming as well, and the glacial anticyclone there was weaker than during the glacial maximum. In North America, spruce had moved northward to the Great Lakes, northern pines were distributed along the entire East Coast to Maine, and oak still grew in the southeastern United States. In Europe, tundra dominated some areas, birch became established in Denmark, and pines moved from the southeast into Poland and central Germany.

After 12 kya lake levels rose as precipitation increased. Crocodiles and hippopotamuses migrated into the Sahara and humans settled near the lakes. Paleoclimate data from the western Arabian Sea show that monsoonal activity strengthened.

9,000 years ago   At 9 kya the summer solar insolation was 8% higher than today, the climate had warmed, and the Laurentide ice sheet had retreated toward Hudson Bay. The glacial anticyclone had weakened in eastern North America, and the Pacific anticyclone had strengthened, sending moist warm air off the Pacific Ocean and Gulf of Mexico into the southwestern and central United States. Lake levels were lower than they were earlier, and these regions were warmer than they are today by 2°C to 4°C. Spruce moved north into Canada. Northern pines moved west and north and were now north of Kentucky and the Carolinas. Oak reached the Great Lakes. Some prairie forbs grew west of the Great Lakes in Canada. In Europe, spruce moved westward from Russia toward Scandinavia. Pines moved into Ireland and became dominant in England and Denmark, and scattered oak came in. Hazel spread throughout western Europe.

6,000 years ago   By 6 kya summer temperatures were 2°C to 4°C higher in both Europe and the United States than they are today. Summer insolation was stronger by 4% and winter insolation was reduced by 4%. This period, from 6000 to 2000 B.C., is known as the **postglacial climatic optimum**, also called the **hypsithermal** or altithermal. Prairie had been established in Illinois, Minnesota, and Canada west of the Great Lakes. Northern pines grew in the Great Lakes region and extended into Canada. Spruce grew as far north as Hudson Bay. After 6 kya the southern limit of spruce in both Europe and North America shifted southward as conditions became cooler once again.

Lake levels in northern Africa remained high 6 kya, but by 3 kya were dropping toward their present-day, very dry condition.

The Younger Dryas Event

On the way to a warmer world, as the North American and Eurasian ice sheets were disintegrating, something strange happened to the climate of the North Atlantic, Greenland, and Europe (Figure 9). Thirteen thousand years ago there was a sudden warming—as much as 4°C—followed by cooling and warming oscillations for about 2000 years. Then suddenly, around 11 kya, the temperature dropped about 6°C in a few hundred years and glacial conditions returned. Ice formed in the North Atlantic as far south as France. The Scandinavian ice sheet and the Scottish Highlands glaciers advanced again. Forests that had colonized postglacial Europe were suddenly taken over by arctic grasses and shrubs. One such plant, the Dryas flower, for which the event was named, is found today only in the arctic and alpine tundra. The Younger Dryas event showed that the global climate system can shift rapidly from interglacial to glacial and back again. Its occurrence is of enormous significance, not only for a better understanding of climate change, but because it may tell us a lot about how plants and animals respond to rapid change. Recent data (see Heusser et al., 1985; Engstrom et al.,

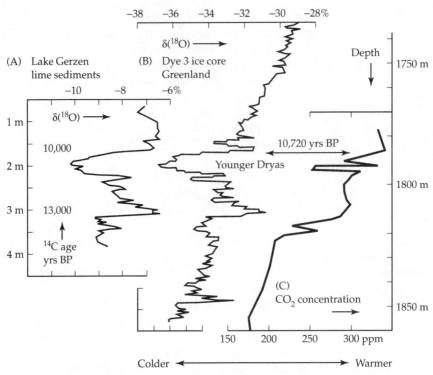

FIGURE 9   Records of the late glacial transition and the Younger Dryas cold event. (A) $\delta^{18}O$ measurements from Lake Gerzen, Switzerland. (B) $\delta^{18}O$ along a 120-meter core from Dye 3, Greenland. (C) Atmospheric CO_2 concentration from gas trapped in polar ice. (After Dansgaard et al., 1989.)

1990; Kudrass et al., 1991) indicate that the Younger Dryas was not just a regional event, but was a global phenomenon between 11 and 10 kya. Although Rind et al. (1986) suggest that palynological and glacial evidence for a worldwide Younger Dryas cooling is not convincing, Overpeck et al. (1989) conclude that "the climate system translated the relatively gradual astronomical forcing, characteristic of the past 18 thousand years, into an inter-related sequence of abrupt climate events over an extensive area of the globe."

Dansgaard et al. (1989) give a precise date for the end of the Younger Dryas—10,720 years ago. Their data suggest that in less than 20 years, the climate in the North Atlantic region became milder and less stormy as a consquence of a rapid retreat of the sea ice cover. A warming of about 7°C occurred in about 50 years—a truly astonishing rate. Their data come from Greenland (Dye 3) ice cores and Swiss lake sediment cores. All parameters studied, including heavy isotopes, chemical trace elements, acidity, and continental dust, reveal the Younger Dryas event and its abrupt termination.

The cause of the Younger Dryas event is still not understood, although several hypotheses have been put forward. Climate oscillations of 2000-year periods are found throughout the interval from 60 to 20 kya in the Greenland ice cores and North Atlantic sediment records. These rapid changes of climate are not likely to be linked solely to the Earth's radiation budget, since that modulates more slowly. It is possible that the atmospheric carbon dioxide concentration had something to do with these oscillations, but it is not likely to have been the only driving force. Bryson (1989) modeled the Milankovitch (orbital) climate cycle of the last 40,000 years and obtained a striking reproduction of the global ice volume when he included volcanic eruptions. In particular, the model reproduced the Younger Dryas event as well as other important details.

Perhaps the answer is to be found in the oceans. Rind et al. (1986) used a GCM into which they incoporated the North Atlantic ocean temperatures of 18 kya, which resembled those prevailing during the Younger Dryas, and also used the orbital changes of solar insolation and land ice distribution of 11 kya. The colder ocean temperatures produced cooling over western and central Europe and also some cooling in extreme eastern North America. Despite the presence of increased land ice and colder oceans, the Younger Dryas summer air temperatures at Northern Hemisphere mid-latitudes in the model were warmer than those of today, due to changes in the orbital parameters (precession) and the subsidence of air (adiabatic warming) at the edge of the ice sheet.

Broecker and Denton (1990) have proposed thermohaline circulation in the oceans, called the "conveyor belt," as an explanation of the Younger Dryas cold interval. Today, warm surface water moves northward in the North Atlantic at depths of up to 800 meters, evaporates along the way, and becomes increasingly saline as well as cooler. The resulting increased density causes it to sink, forming cold bottom water which then moves southward. According to Broecker and Denton, this process releases an enormous amount of heat into the atmosphere over the North Atlantic, and it is this phenomenon that accounts for the mild winters of Western Europe—not the Gulf Stream, which terminates further south. Broecker and Denton write:

The magnitude of the vertical circulation is also immense, averaging 20 times the combined flow of all the world's rivers. Indeed, much of the deep water in the world's oceans ultimately originates here. From its source the water floods the deep Atlantic, curves around the southern tip of Africa and joins the deep current that circles Antarctica and distributes deep water to the other oceans.

Broecker and Denton assert this "conveyor belt" has about a 2000-year cycle that switches between "on" and "off" modes. The last deglaciation had four intervals of increased meltwater discharge into the North Atlantic with an approximately 2000-year cycle. A large volume of fresh water entering the ocean could alter its salinity, switching off the conveyor. Broecker and Denton propose that a massive influx of fresh meltwater off the North American glacier flowing into the North Atlantic blocked the conveyer and precipitated the Younger Dryas cold. Because of the cold, a lobe of ice advanced across the flow from Lake Agassiz (the Great Lakes) and diverted the meltwater down the Mississippi watershed. Then the conveyer belt reactivated itself and Europe warmed again. The consequences to climate were strongest on the periphery of the North Atlantic. Broecker and Denton suggest that carbon dioxide, methane, and dust also played a role. The ice age atmosphere was very dusty, which could have contributed to the cooling.

Broecker and Denton suggest that there may be several stable states to the atmosphere and that transitions between two of these, glacial and interglacial, characterized it in the past. Upsetting the deep water flow in the North Atlantic could be a mechanism for flipping from the interglacial to the glacial state. This hypothesis is disputed by Shackleton (1989) based on a paper by Fairbanks (1989) with new evidence of glacial melting rates before and during the Younger Dryas event.

The Younger Dryas event is discussed again in later chapters with respect to the response of organisms to climate change.

The Little Ice Age

Following a period of warmth referred to as the Middle Ages warm epoch, which lasted off and on from 1100 to 1400, there set in a period with frequent long, severe winters and short, wet summers. This is known as the Little Ice Age and lasted from 1430 to 1850 (see Figure 6). The extent of the snow and ice on land and sea attained a maximum greater than at any time since the last major ice sheet. It appears that the atmospheric circulation became more meridional than before or afterwards, meaning that deeper north-to-south movement of air masses occurred, in contrast to the more common strong west-to-east flow referred to as zonal.

During the Little Ice Age many glaciers in Alaska, Scandinavia, and the Alps advanced in synchrony to nearly maximum positions. Expansion of the Arctic sea ice into the North Atlantic cut off the transport of supplies between Europe and the Norse colony that Erik the Red had established in southwest Greenland in about 978. The result was that the food supply and nourishment of the colony diminished, and the colonists eventually perished. In Iceland, grain that had grown there for centuries could no longer survive. There was great crop failure and famine in Europe in 1433 to 1438. Many European villages that had been settled during the Middle Ages were abandoned between 1400 and 1480. There was suffering in Europe as early as 1310 to 1360 and some abandonment of villages then. The climate varied considerably toward the end of the Middle Ages. The Gulf Stream took a more southerly course across the Atlantic. It was as if the countries of Europe had moved 300 miles to the north.

The Little Ice Age is a time of enormous interest to climatologists. It was suggested by Eddy (1977) that the Little Ice Age was caused by variations in solar insolation resulting from a minimum in the sunspot cycle, the Maunder Minimum (1645 to 1715). However, it turns out that this period was not the coldest part of the Little Ice Age. The coldest times came in the early 1600s and early 1800s. Robock (1979) showed that sunspots had little influence on the climate of the Little Ice Age and that volcanic dust had a significant effect. More recently, Thompson et al. (1986) reported ice core results from the Quelccaya ice cap in Peru, which showed the Little Ice Age to have been a global event. The striking thing about the Little Ice Age is that 5- and 10-year average Northern Hemisphere temperatures were only about 1°C less than temperatures earlier or later. Now oceanographers are thinking about how the dynamics of North Atlantic circulation may have produced the Little Ice Age. But the real cause of this cold period has not yet been resolved satisfactorily. See Grove (1988), Ladurie (1971), and Lamb (1977) for further information.

Changing Communities

The Earth's changing climate has helped to produce vast diversity among the organisms living on the planet. As the climate changed, some species adapted while others went extinct. Those species that adapted usually experienced major changes in the extent of their geographical ranges. The distribution of plant species is of particular importance because plants are **primary producers**—capable of transforming energy from sunlight and inorganic chemicals. Animals obtain energy by feeding on plants, or on other organisms that feed on plants. Without the Earth's vegetation, animal life would not exist.

In addition to supplying energy, plants are a major feature of the diverse habitats to which different organisms have adapted. Picture a New England forest, the Everglades of Florida, and a Canadian prairie— all vastly and obviously different in their vegetation and the life forms that survive among the vegetation. The next three chapters focus on plants and on the vegetation communities that are necessary to our survival and to the diversity of organisms on the Earth.

Plant Physiognomy and Physiology

THERE HAS LONG been an interest in the relationship between climate and vegetation distribution. If a clear relationship between vegetation and climate can be established, then the one can be used to infer the other, and vice versa. Furthermore, the relationship can be used to project the future condition of vegetation under climate change.

On global or continental scales, climate is a primary determinant of the distribution of vegetation types. Phylogenetically unrelated plants growing in similar climates in different parts of the world have evolved growth forms that look much alike. This is known as *convergent evolution*. Geographers of the nineteenth century recognized that vegetation types reflected the climate of the region by their manner of growth and their appearance, or physiognomy. Richard Hines, about 1840, was among the first to classify climates and relate them to the distribution of plants and animals.

As plant physiology became better understood, scientists began to connect the physiological requirements of plants with the climate regimes in which they grew. In 1874 the botanist DeCandolle established a climate classification system based on plant physiognomy and physiology. Wladimir Köppen published papers in 1900, 1918, and 1936 in which he developed a more modern and quantitative elaboration of the DeCandolle system. He used vegetational distributions to classify climate. The Köppen classification has been refined by Guetter and Kutzbach (1990). Holdridge (1947) related potential natural vegetation distributions to annual precipitation and growing-season temperatures. Holdridge's classification is discussed further in Chapter 5. See Gates, 1972 for a description of other climate classification schemes.

Ecologists divide the Earth's ecosystems into **biomes**—regions characterized by their distinctive vegetation. Climate is a definitive factor of

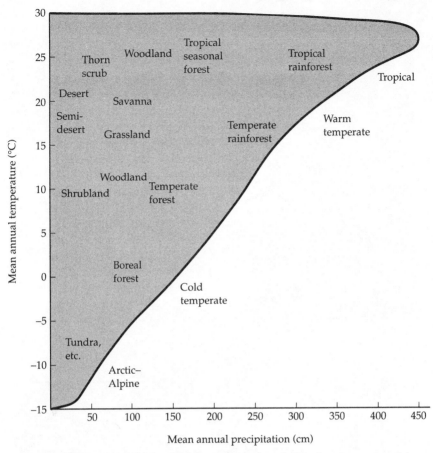

FIGURE 1 The pattern of world biome types in relation to mean annual temperature and precipitation.

a given biome; biome types can be related to the mean annual temperature and mean annual precipitation where they occur, as shown in Figure 1. It might seem that this would tell us much of what we need to know concerning plants and climates. Unfortunately, annual averages are just that, and give us little indication of the importance of **seasonality**—the variation of temperature throughout the year—in determining vegetation assemblages. For example, where winters are very cold and summers are short, cool, and dry, one finds tundra; where winters are very cold and summers are warm with moderate rainfall, one finds boreal forest; where winters are cool and summers are warm with precipitation throughout the year, one finds broadleaved deciduous forest (Figure 2).

(A)

(C)

(B)

FIGURE 2 Three biomes whose positions will alter greatly under predicted climate changes. (A) Tundra. This aerial view shows the absence of trees and the prevalence of small seasonal ponds. The bumpy landscape is due to frost action. A herd of reindeer appears toward the center of the photograph. (B) A boreal forest in northern Idaho. The predominant trees in this photograph are spruce. (C) A temperate deciduous forest in Indiana. Oaks, hickories, and other broadleaved hardwood trees abound. (Photographs courtesy of U.S. Forest Service.)

Plant Demand for Water and Energy

The difficulty with climate classification schemes is well summarized by Stephenson (1990):

Most studies of vegetation physiognomy and climate have described climate in terms of measures related to annual energy supply (such as temperature, potential evapotranspiration, or radiation), annual water supply (precipitation), or their ratios. Unfortunately, these measures cannot distinguish between climates similar in annual energy and water supplies but different in the seasonal timing of the two and, therefore, implicitly assume that the effects of energy and water are independent. As sensed by plants, however, the effects are not independent. For a plant to use energy for growth, water must be available; otherwise, the energy acts only to heat and stress the plant. Similarly, for a plant to use water for growth, energy must be available; otherwise, water simply percolates through the soil or runs off, unused. The effects of climate on plants, therefore, are determined by the interactions of energy and water.

Stephenson showed that the distribution of North American plant types is more highly correlated with water balance (annual actual evapotranspiration and annual water deficit) than with annual temperature and precipitation. He defines *deficit* as the evaporative demand that is not met by the available water supply. He defines *actual evapotranspiration* as a measure of the simultaneous availability of biologically usable energy and biologically usable water. It is "the evaporative water loss from a site covered by a standard crop, given the prevailing water availability. Actual evapotranspiration equals available water or potential evapotranspiration, whichever is less." Potential evapotranspiration always equals actual evapotranspiration plus deficit. Seasonality directly affects the calculation of these terms.

Stephenson's arrangement is summarized in Figure 3. Diagonal lines across the diagram represent isolines of potential evapotranspiration. A line connecting actual evapotranspiration of 1000 mm and a deficit of 1000 mm represents a transect of increasing aridity (decreasing actual evapotranspiration and increasing deficit) at approximately constant energy supply (potential evapotranspiration = 1000 mm). This would represent an east–west transect along 40° north latitude in North America across a sequence of deciduous forest, tallgrass and shortgrass prairie, and cold desert-shrub in the far west. It is notable that forest types are distributed along the annual actual evapotranspiration axis and that all are limited by annual water deficits greater than 400 mm. Furthermore, the distinction between conifer and deciduous forest does not depend upon the annual water deficit, or drought.

There are problems with all of these classification methods, largely because both climate and vegetation distributions are collective concepts that bring together many parameters. Plants respond to climate as individual species, largely based on their physiological requirements. As long-term climate changes occur, biomes will re-form with different plant associations. However, for broad-scale studies of ecosystems, it is useful to combine physiological understanding with more general physiognomy (e.g., leaf shape and size, leaf area index, canopy structure, life form, and seasonality). Much depends on whether one is trying to understand the distributions of ecosystems and biomes formed under the influence of long-term steady-state climates, or the dynamics of ongoing vegetation change. The models of the process, known as **forest gap models** (described in Chapter 5), are dynamic and capable of bridging this dichotomy. A new scheme to model large-scale vegetation patterns with steady-state climate is given by Prentice et al. (1992b).

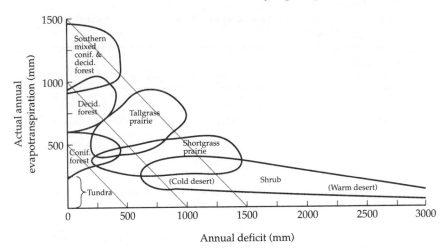

FIGURE 3 **The distribution of major North American plant assemblages in relation to annual actual evapotranspiration and annual precipitation deficit. (From Stephenson, 1990.)**

Although climate is the major factor, soil, topography, fires, and human and animal disturbance all have a secondary influence on the distribution of vegetation.

Plant Physiognomy

Classifications of climate and vegetation have become increasingly more quantitative and precise. Woodward (1992) used physiognomy to classify vegetation worldwide according to annual minimum temperatures and growing degree-day (GDD) totals, which he calls day–degree totals (Tables 1 and 2). Plant species differ greatly in their ability to tolerate either cold or heat. Chilling-sensitive plants are damaged by temperatures lower than 10°C. This category includes most tropical plants. Freezing-sensitive plants can tolerate low temperatures, but are damaged when ice crystals form within their tissues. In Table 1 Woodward has distinguished those that cannot survive below 0°C from those that can survive temperatures down to about −15°C. The difference is that the latter group can, by adjusting the osmotic concentration of substances in their cell sap and protoplasm, reduce the freezing point to −15°C. Other plants are frost-tolerant. These survive by withdrawing water from their cells so that ice crystals cannot form within them. This process also concentrates the solutes in sap and protoplasm, thereby reducing the freezing point. Frost-tolerant plants include the broad-leaved deciduous species, which have a lower temperature limit of

TABLE 1 **Annual minimum temperatures for expected dominant physiognomies.**

Minimum temperature (°C)	Physiognomy
>10	Broadleaved, evergreen, chilling sensitive
0 to 10	Broadleaved, evergreen, chilling resistant
−15 to 0	Broadleaved, evergreen, frost resistant
−40 to −15	Broadleaved, deciduous
<−40	Some broadleaved, but mostly evergreen and deciduous needles

(From Woodward, 1992.)

−40°C. Finally there are the cold-tolerant plants, including some broad-leaved, but mostly evergreen and deciduous needle-leaved plants, that can tolerate almost any amount of cold. The physiological processes of adjustment to low temperatures are known as **cold hardening**.

Woodward defines growing degree-day (GDD) units as the product of the length of the growing season (the time in days for which the mean temperature exceeds 0°C) and the mean temperature over this period. Various definitions of GDD have been used; these are discussed further in a later chapter. The GDD totals needed by species of differing physiognomy to complete their vegetative and reproductive life cycles are given in Table 2. *Koenigia islandica*, a tundra plant, requires a total of only 700 GDD to develop from seed to seed, whereas the lime tree, *Tilia cordata*, requires a minimum of 2000 GDD to complete its reproductive development. Trees in the tropical rainforest may require almost

TABLE 2 **Growing degree-day totals for expected dominant physiognomies.**

Physiognomy	Growing degree-days (GDD)	
	Reproductive	Vegetative
Broadleaved deciduous	2800–2100	2100–1700
Evergreen coniferous	2300–1500	1500–900
Deciduous coniferous	1900–1100	1100–700
Tundra	1600–700	700–200

(From Woodward, 1992.)

10,000 GDD to complete their reproductive development. Woodward is correct in pointing out that he may be overestimating the GDD threshold here, since he uses 0°C instead of the more likely 10°C growing season threshold for tropical trees. Most researchers use 5°C even for boreal trees, as we shall see later.

A cool climate limits the number of species that can grow, but as the minimum temperature increases in the equatorial direction, warmer-adapted species will outcompete the cool-climate vegetation. This is an important concept to keep in mind when we discuss the effects of global warming in Chapter 5.

Plant Physiology

Temperature

We have seen that temperature limits the distribution of vegetation and, in particular, of forests. Temperature affects photosynthesis, respiration, growth, reproduction, and water usage among other things.

A plant's rate of net photosynthesis increases with temperature until a broad optimal temperature range is reached; then it drops off at higher temperatures (Figure 4). This pattern is the result of an increase in gross photosynthesis at low temperatures and a rapid rise of respiration rate at high temperatures. Plants adapted to cool temperatures have optima of about 10°C and do not photosynthesize above about 25°C. Temperate-zone plants generally have optima between 15°C and 30°C. Plants adapted to warm temperatures may have optima as high as 40°C. Larcher (1975) wrote, "The rates of photosynthesis and respiration adapt to the temperature prevailing at a given time." We know that cold-adapted plants can shift their photosynthetic optimum toward higher temperatures when they are grown under warmer conditions. This results from a shift in the enzyme systems that affect both photosynthesis and respiration. Even seasonal shifts of temperature optima are found in some evergreen plants. These shifts in temperature optima may be only a few degrees; nevertheless, they are effective means of adjustment to changing climates.

The increase of respiration rates with temperature are of particular interest to our concern about the effects of global warming. As an approximation, a 10°C increase in temperature doubles a plant's rate of respiration. Since respiration releases CO_2 to the atmosphere, this represents a positive feedback to, and thus a serious factor in, global warming. Not only will plants respire faster with increasing temperatures, but soil respiration by microbes will also increase considerably and will release additional CO_2. Plant litter breaks down 33% more rapidly at 17°C than at 14°C.

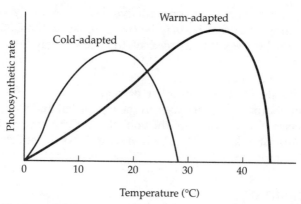

FIGURE 4   **Photosynthetic rate as a function of temperature for a warm-adapted and a cold-adapted plant. (From Gates, 1985a.)**

Warmer temperatures increase the growth rates of most plants, particularly near the center of their climate ranges, provided water and nutrients are available, although warming beyond a certain point will result in reduced rates of growth and even dieback. If the mean July temperature increases from 14°C to 17°C, the radial growth rate of Scots pine increases by as much as 54%, according to Cannell et al. (1989).

Plants in continental regions, where there is great seasonality, develop best if nights are about 10°C to 15°C cooler than days, according to Went (1961). Maritime species, with more equable climates, prefer a difference of only 5°C to 10°C. Some tropical plants, such as sugarcane and peanuts, do not need cool nights, and some, like the African violet *Saintpaulia*, prefer nights warmer than days. Many tree species, such as the Douglas fir, require chilling in the winter in order to grow well the following summer. High temperatures in early winter may be detrimental to the growth of oaks because they increase the trees' rates of respiration and prevent sufficient chilling. Many deciduous plants of temperate regions require the chilling of buds in order to break winter dormancy, and without this chilling they will die. The whole subject of chilling requirements for breaking dormancy—called vernalization—is extremely complex and far from well understood. However, when we consider the possible effects of global warming, it important to realize what the loss of winter chilling may mean.

Seed Germination

Seed germination involves metabolic activity and therefore has a temperature-dependent function (see Lang, 1965). Some seeds require a period of dormancy in a certain temperature range for germination to occur, and others do not. As with other plant processes, the rate of germination increases with temperature until an optimum range is reached, then declines at higher temperatures. Tropical plant seeds germinate best between 15°C and 30°C, temperate-zone seeds between 8°C and 25°C, and alpine plant seeds between 5°C and 30°C. For most seeds, the highest germination success occurs not with constant temperature but with a fluctuating temperature regime. The diurnal (daily) cycle will usually satisfy this requirement. Many cultivated plants, particularly those in temperate zones, have a wide germination temperature range, extending from the freezing point to 30°C or above. The seeds of some weeds will germinate between 2°C and 40°C. Then there are some rock-garden plants and biennial desert plants whose seeds will germinate only within a narrow temperature range around 5°C. It is also known that seeds produced by a plant grown under warm conditions will have an optimum germination temperature higher than that of seeds from a plant of the same species grown under cool conditions. Many seeds also depend on soil moisture conditions for germination. Some seeds contain an inhibitor substance that prohibits germination until rain dilutes it out (see Went, 1957).

Some seeds are especially sensitive to light. "Pioneer" species that begin the successional process have seeds that sprout and grow rapidly under bright light, although they may be affected by soil moisture conditions. Seeds of late-successional species, such as the American beech and sugar maple, are unaffected by light and so can grow in areas shaded by the plants that grew before them.

Cold Hardening

As mentioned earlier, plants may adapt to cold and prevent freezing by various cold-hardening mechanisms. The ability of plants to undergo cold hardening is species-specific and varies during the year. The ability is minimal during the growing season, develops gradually during the autumn, and reaches a maximum during the winter. If there is a warm period of a week or more during the winter, some of the cold hardening will be reduced and will not be entirely recovered for the remainder of the winter. This means the plant will be particularly vulnerable to cold spells after that. (This problem is discussed further later in this chapter; see the section on "Causes of Forest Decline.")

Examples of cold hardening by cold-adapted tree species in the summer versus the winter are given by Larcher (1980). White fir organs are cold-hardened during summer and during winter at the following temperatures: needles, −3°C and −35°C; twigs, −5°C and −35°C or lower; bole, −18°C; and roots, −3°C and −20°C. Root tips may freeze at temperatures below −2°C, and the permanent roots may freeze below

−5°C to −20°C, depending on the tree species. The relative lack of cold hardening by roots shows why the insulating value of snow cover is especially important (see Gates, 1980).

Heat Resistance

Plants develop resistance to heat stress on an annual cycle, although the changes are not as great as those due to cold. Damage to a plant by heat is proportional to the product of temperature and time. A plant's susceptibility to high temperatures depends upon many factors, including its age, the season of the year, vigor of growth, previous temperature history, light intensity, and genetic (inherited) tolerance (see Gates, 1980 for a detailed discussion). Different parts of a plant develop differing degrees of tolerance to heat. Sometimes the uppermost leaves become the most heat resistant, while for other plant species it may be the leaves in the middle of the canopy, and for others, the leaves at the base of the plant. Evergreens and shrubs from cold winter areas may develop heat injury in summer at about 44°C; temperate-zone trees and shrubs usually show damage at 50°C or above; some subtropical woody plants and succulents at 50°C to 55°C; and tropical trees between 45°C and 55°C. Up to a certain temperature, plant cells are constantly repairing whatever damage may occur; but above this temperature, the heat damage is irreversible. Most often, irreversibility occurs above 50°C.

Water Use and Temperature

The water balance of plants is an important topic in plant physiology that we will not deal with in detail here. Stands of vegetation transpire from 50% to 80% of the annual precipitation. Heath and tundra plants transpire about 15% to 25% of the annual precipitation. An increase in temperature increases evapotranspiration rates because as leaf temperature increases, the water vapor pressure inside the leaf increases and the vapor pressure deficit (VPD) increases. As VPD increases, the gradient driving water vapor out of the leaf increases and the plant loses water more rapidly. Vapor pressure deficit also increases as the atmosphere becomes drier.

The energy budget of a leaf involves the exchange of energy by radiation, convection, and transpiration. (See Gates, 1980, and Gates and Papian, 1971 for detailed discussions of energy exchange.) A leaf assumes a temperature and transpiration rate that balances its energy budget. If the leaf's temperature increases, there is an adjustment of each quantity in the energy exchange process. I have used some of the calculations in the above-mentioned references to calculate the possible increase of transpiration rate with temperature. There are so many variables involved that I shall omit the details. I find, for a common-sized leaf with normal diffusion resistance, that with a 10°C temperature increase, the transpiration rate will increase about 20%. Cannell et al. (1989) get a 5% increase with constant VPD for a 3°C temperature increase for forest trees. With increasing VPD (drier air, for example) they get as high as a 122% increase. They feel, however, that the effects of changing precipitation patterns may be of greater importance to plants than changes in their rates of water usage. Nevertheless, increased transpiration rates with increasing temperature may dry soils more rapidly and produce an added stress on many forest trees and even on some herbaceous plants.

Tilia cordata and Temperature

The lime tree, *Tilia cordata*, is distributed across much of Europe. Its northern limit in England and Scandinavia parallels the mean July 19°C isotherm. One of the nicest pieces of detective work in plant ecology was done by Professor Donald Pigott and his students while tracking down the precise reason for this temperature relationship. Their study involved not only observational work and measurements in the field, but detailed cytological studies in the laboratory and physiological measurements as well (see Pigott, 1981, and Pigott and Huntley, 1981).

The lime tree reached the Lake District in the north of England around 7000 years ago. Today, a few relict trees are found in the north of England, and these are reproducing only vegetatively. The species requires 2000 GDD or more to produce seeds by sexual reproduction. Its present northwest limit was established around 5000 years ago when conditions had cooled considerably from the warmest times of the hypsithermal. Remember in Chapter 2 we observed that after 6 kya the southern limit of spruce in both Europe and North America shifted southward as conditions became cooler once again. Pigott and his students found that, for *Tilia cordata* to reproduce, not only must flowers develop, but the pollen must germinate and then be transferred through a pollen tube to the ovary for fertilization. They found that the pollen will not germinate at temperatures of 15°C or below. The optimum temperature for germination is 17°C to 22°C; above that germination success decreases until it reaches zero at 35°C. However, it turned out that the growth of the pollen tube is also temperature-dependent. The rate of tube growth is at its maximum around 20°C to 25°C and diminishes greatly at higher and lower temperatures. In fact, the best estimate of the temperature above which the extension of the pollen tube becomes rapid is 19°C.

This study of the relationship between temperature and the reproductive success of a forest tree is the finest piece of work of this kind ever done and should be done for many other tree species. It is clear from this study that even a 1°C or 2°C shift in the mean July temperature, and particularly in the frequency of maximum temperatures above 19°C, will have a dramatic impact on the distribution of the lime tree. In many places where the tree grows today, but is not fertile, it will become so with warmer temperatures in the north.

Effects of Increased Carbon Dioxide

All plants should increase their photosynthetic rates with an increase in the atmospheric concentration of carbon dioxide. The photosynthetic rates of many plants, compared with their rates at the present-day concentration of CO_2 (approximately 354 ppm), will increase by 50% to 75% with a doubling of the CO_2 concentration. Trees growing in northern Michigan were shown to increase their photosynthetic rates by 50% with a doubling of the current CO_2 concentration (Gates, 1985b). For agricultural crops, an increase in growth of between 50% and 70% translates into an increase in biomass yield of about 40% to 60%, and into a seed yield increase of 30% to 50%. However, under normal agricultural practices the numbers may be somewhat lower because of limitations caused by nutrient or water availability. Ironically, no one has shown unambiguously that plants have had increased rates of growth in the natural world due to the increase in CO_2 concentration from 275 ppm to 354 ppm over the last 120 years. However, practically all greenhouse and growth chamber experiments have shown increased photosynthetic and growth rates with increasing levels of atmospheric CO_2.

Bazzaz (1990) gives an excellent review of the response of natural ecosystems to rising levels of atmospheric CO_2. He concludes that C_3 plants are more responsive to increased CO_2 levels than are C_4 plants;* that photosynthesis is enhanced by CO_2 increase, but may decline with time; that the response to CO_2 is more pronounced under higher levels of nutrients, water, and light; different species, even in the same plant community, may differ in their responses to CO_2.

Other plant responses to CO_2 enrichment include a change in the respiration rate, a decrease in the transpiration rate (linked to a reduction in stomatal opening), increased tolerance to atmospheric pollutants, increased leaf area, increased root-to-shoot ratio, earlier flowering, an increase in the size of fruit, and other responses too technical to merit discussion here. These individual plant responses effect ecosystem

---

*C_3 and C_4 plants differ in the biochemistry of their photosynthetic processes; C_3 plants assimilate carbon dioxide into a 3-carbon chemical compound, whereas C_4 plants assimilate carbon dioxide into one of several 4-carbon compounds. C_4 plants often live in warm environments where water is scarce or only sporadically available.

changes through the entire food chain. Leaves may be larger when grown in enriched amounts of CO_2, but if the nitrogen supply is not also increased, they may have less nitrogen per unit of weight, and insects will have to eat more leaf volume to get the nutrition they require. Competition among plant species will shift. Water use efficiency is defined as the ratio of the assimilation rate to the transpiration rate. Generally, the water use efficiency of plants increases with increasing amounts of CO_2, but if the climate gets warmer, it could decrease instead. (More information may be found in Lemon, 1984, and in Gates et al., 1983.) Woodward (1987b) reported a reduction of stomate size and a decrease in leaf conductance* in greenhouse-grown plants under increased levels of CO_2. He confirmed this by looking at the stomata of herbarium specimens of plants collected at intervals over the last 100 years. During this time, CO_2 concentrations have increased about 28%.

Causes of Forest Decline

The decline of forests during recent years has been the subject of worldwide interest. Although the spectacular destruction of the tropical rainforests has received massive public attention, the decline of forests in Europe and North America has been intensely studied. This decline has been attributed to acid deposition in much of Europe and in the Appalachian Mountains of the eastern United States. (The numerous factors affecting forest decline have been reviewed by Klein and Perkins, 1988.) But some types of forest decline that have occurred episodically over the past 60 years in eastern Canada and the northeastern United States are clearly not just the result of air pollution. Auclair (1989) has reviewed these decline events and related them to climate.

Dieback events for forest trees in southeastern Canada have included diebacks of ash species in 1925, yellow and white birch in 1932, American beech in 1934, sugar maple in 1978, and much sugar maple and birch during the 1980s. The combined pattern of these diebacks is shown in Figure 5. There is a strong coincidence of diebacks with warm periods in the 5-year running mean temperature of the Northern Hemisphere between latitudes 64.2° and 90° north, with recoveries during cooler times. This does not by itself implicate warmer conditions as the cause of dieback. However, local weather records in the region of dieback clearly indicated extreme weather events during some of the years of maximum dieback. These extreme weather patterns included snowless

---

*Stomates or stomata are the openings in the leaf surface through which gas exchange and transpiration take place. Leaf conductance—the readiness of the plant to lose water or take up CO_2 through its leaves—decreases as the size of these structures decreases.

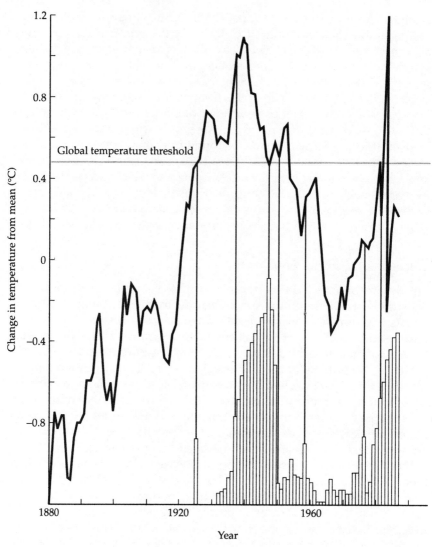

FIGURE 5 The total level of forest decline in ash, birch, and maple shown in relation to the trend of mean air temperatures north of 64.2° north latitude (shown as a five-year running mean). The approximate global temperature threshold for onset and recovery of regional forest decline is shown by the horizontal line. (From Auclair, 1989.)

winters, winter thaws followed by intense cold snaps, spring frost, spring and summer drought, and heat stress.

All of the tree species undergoing decline are deciduous hardwoods characterized by shallow roots, particularly the birches. During most winters there is typically snowpack of sufficient depth to insulate these roots against the penetration of intense cold into the soil. This protection is particularly important for feeder roots, which take up nutrients from litter and humus near the soil surface. In some species these roots are mycorrhizal and carry on nitrogen fixation. When warm periods occur in the winter, snow depth decreases; when this is followed by intense cold, frost penetrates to the feeder roots, and even deeper. Although all plant parts become cold-hardened during the winter, some of this hardening is lost during warm intervals, so the feeder roots become vulnerable to cold. Furthermore, freeze–thaw cycles can physically break the feeder roots. When the feeder roots, and even some of the fine roots below them, are damaged, the tree cannot take up sufficient nutrients and water during the following growing season. Nutrient deficiency, particularly of phosphorous, and water deficits characterize many declining trees.

Dieback occurs most severely in dominant or co-dominant older trees, which have a high shoot-to-root ratio relative to understory trees. Trees with the highest radial growth rates are the first to be affected; trees that remain healthy during dieback are those with the slowest annual growth rates. Dieback is most severe for trees growing on crests or in bottomlands. The upper crest sites are more vulnerable to wind, heat, and drought. In bottomlands, roots are concentrated near the surface because of soil waterlogging underneath them; in addition, bottomlands form natural frost pockets. Trees that are stressed are most vulnerable to disease, pathogens, and insects. I have seen many instances of birch dieback in northern Michigan during recent warm summers when the surface soil has become extremely dry. In every instance, the birch trees were infested with the birch borer. Pomerleau and Lortie (1962) showed that shallow-rooted white birch trees were particularly susceptible to dieback, compared with those with deeper root systems. Leaphart and Stage (1971) showed that drought brought on the pole blight disease in western white pine. Staley (1965) reported an extensive study of decline and mortality in red and scarlet oaks. He suggested that the primary cause of decline was defoliation by the leaf roller which in turn reduced the trees' carbohydrate production and growth. In this instance, drought was considered a secondary cause of dieback.

Pastor and Post (1986) used a combination of computer modeling and field data to describe the interactions of climate, soil moisture, and nitrogen cycling with the processes of reproduction, growth, and death in forest trees. They presented a table giving the drought tolerances of various tree species. Some oak speces are extremely drought tolerant while other oaks are very intolerant. Birches, as a general rule, are quite intolerant to drought. Sugar maple was one of the most drought intolerant species in the entire list.

Auclair et al. (1990) have noted the dieback of trees in northwest British Columbia and southeastern Alaska, where comparatively large climate change has been documented over the past century by Wigley et al. (1980). Yellow cedar has shown a decline of radial growth beginning in the 1880s and coinciding with a marked increase in annual air temperatures in Alaska. Auclair et al. have documented the decline of five other tree species in the Pacific Northwest and point out that this decline has occurred in the absence of any air pollution.

In addition to root mortality, another cause of forest decline is cavitation. Hardwood trees transfer water from their roots to their crowns through long conducting vessels known as tracheids, which are often several centimeters or even meters long. As evapotranspiration from the crown takes place, the difference in water potential between the crown and the roots causes tension to develop in the water column. If air gets into this water column, embolisms develop that break the water column and interfere with the transport of water to the crown. This process is known as **cavitation**. In a sudden freeze, dissolved gases come out of solution, causing air bubbles to form, and cavitation occurs.

Trees that suffer severe cavitation cannot transport enough nutrients and water to the crown during the growing season. In sugar maples under cold, sunny conditions, with the water column under tension, cavitation can affect as many as 69% of the conducting vessels in the boles and 89% of those in the twigs. Tyree and Sperry (1988) and Sperry and Tyree (1988) described the detection of cavitation by means of an acoustical device that allowed them to listen to cavitation taking place. All the species they studied operate near the point of catastrophic failure in their water transport systems due to dynamic water stress. Furthermore, they found that "plants are designed hydraulically to sacrifice highly vulnerable minor branches and thus improve the water balance of the remaining parts."

Fire

Fires and vegetation have always worked in synchrony. Fire controls the spread of some types of vegetation and promotes the spread of other species. Fire controls the birth of some plants and is responsible for the death of others. Insects and some plant pathogens are suppressed by fire, and animal populations are affected by it. Fire controls the release of carbon and nutrients in the vegetation and in the organic layers of the soil. Tallgrass prairie is perpetuated by fire suppressing the advance of woody vegetation. The patchiness of the boreal forest results from the frequency of fires.

A warmer, drier climate is likely to increase the frequency of fires. Some forest trees germinate and grow well following fires. The resinous cones of jack pine and lodgepole pine open only when heated by fire; their seeds fall onto the ground and subsequently germinate. Fire has also been important for the establishment of oak forests in the eastern United States. According to Abrams (1992), oak savannas became established in south-central Wisconsin between 6500 and 3500 years ago, when conditions were warm and dry and fires were frequent. Then the climate became cooler and wetter, with fewer fires, and oak savannas became closed oak forests. Oaks are replaced in succession by hardwood species such as sugar maple, red maple, and black cherry. Fires suppress these hardwood species and allow the oaks to form a closed canopy. Oak seedlings have a very low tolerance to shade and therefore do not grow to maturity under a closed, or nearly closed, canopy; this accounts for the lack of replacement by oak in many old-growth oak stands.

Phenology

Phenology is the study of periodic biological events and their relationship to seasonal climate changes. For animals, such events include mating and reproduction, changes in the thermal quality of their fur, metabolic changes, behavioral traits such as migration, and many others. For plants, phenological events include growth, bud initiation, bud burst, leaf development, flowering, seed development, and leaf senescence. For crop plants, they include time of planting, time of harvest, and length of time for the ripening of fruit.

By observing the timing of vegetative processes, one can trace a continuous pattern of climate across a landscape. A careful study of the timing of phenological events for a particular plant or animal may be of considerable value when interpreting climate change over a period of many years. Leafing out of vegetation may start earlier in the spring, and in the Northern Hemisphere it will progress northward more rapidly with a long-term warming of the climate. Birds may migrate northward earlier and remain in high latitudes longer in warmer times. For example, by knowing the dates of the flowering of cherry trees in China over time, we can infer the climate over hundreds of years for which meteorological records were not available. The dates for wine harvests, which are well known in Europe going back to medieval times, can also provide information about climate changes. (See Ladurie, 1971 for a fine description of phenology and climate.)

There is a well-known bioclimatic law, called Hopkins' law, concerning the rate at which time of occurrence of a phenological event moves across the landscapes. Hopkins (1918) states that:

. . . the time of occurrence of a given periodical event in life activity in temperate North America is at the general average rate of 4 days to each 1 degree of latitude, 5 degrees of longitude, 400 feet of altitude, later northward, eastward and upward in the spring and early summer, and the reverse in late summer and autumn.

Caprio (1967) describes the use of lilac flowering as a climate indicator in the western United States. He shows the date of lilac flowering to be an indicator of flowering dates for several other plant species that bloom somewhat later in the spring. He finds that Hopkins' law does not hold strictly everywhere and that there are important local variations from it. These variations are produced by such things as winds off the mountains, (e.g., the foehn of alpine Europe or the chinook of northwestern North America), temperature inversions in broad valleys, and coastal temperature effects from ocean currents. In some parts of California the lilac blooms move upward in elevation at about 100 feet per day, in agreement with Hopkins' law, while the temperature or thermal wave only moves at 65 feet per day. In the western Great Plains, the flowering moves upward more rapidly than 100 feet per day. Caprio suggests that something else is going on here, that possibly a change in radiation intensity, or perhaps an integrated thermal influence such as degree-days, is driving the flowering.

Hulbert (1963) published the dates of first flowering for 132 plant species observed by Frank C. Gates at Manhattan, Kansas from 1926 through 1955. The records show a great variation in time of flowering from year to year. The observations show that the range between the earliest and latest dates of flowering is much greater (60 days) for those species that flower early than for those that flower later (26 days). The flowering dates indicate something of the climate character of individual years. For example, 1944 had a warm period in January during which American elm and silver maple bloomed; but the spring was unusually cool and other species did not flower until much later than normal.

There is much to be learned about phenology and there is a great deal it may tell us about climate change in the future. The study of phenological events, including animal migrations, may give us a continuum of information across the landscape concerning climate change. The onset of spring, generally marked by the start of vegetative growth, may be modified by climate change. Schwartz (1990) has used phenological data for lilac and honeysuckle from the U.S. Department of Agriculture regional phenology network in the eastern and central United States as a "spring index" marker, which he correlated with synoptic meteorological data derived from daily maximum and minimum temperatures. He modeled the time of first leaf appearance in the spring, since it represents the start of photosynthetic activity and may be related to changes in surface albedo and transpiration rates. This measure of a spring index correlated well, as a first approximation, with the synoptic weather data.

Once and Future Vegetation Patterns

Sustained climate—warm or cold, moist or dry—determines the distribution of vegetation types throughout the world, and seasonality may also play a significant role. We can see vegetation changes taking place on relatively short time scales, such as over a single winter or summer, and we know that these types of changes have occurred throughout the past. Although it is difficult to integrate short-term changes into longer sequences of persistent climate patterns, tree migration and vegetation changes that are the result of prolonged climate forcing may be inferred from the analysis of pollen deposits in the lakes of a region. The next chapter takes us on a journey through time and across the landscapes of many past millenia, teaching lessons from which we may infer future vegetational patterns as the climate changes again.

C H A P T E R   F O U R

Past Vegetational Changes

JUST AS IT IS important to understand the climates of the past in order to better predict those of the future, it is essential to understand the past distribution of vegetation communities in order to anticipate the changes to be expected in a warming world.

The degree of stability of plant communities through time and space determines how they will respond to climate changes. **Stability** means the tendency for vegetation to return to a steady-state condition after being disturbed. Watts (1979), studying pollen sequences from the central Appalachian Mountains, concluded there was considerable vegetation stability during the full-glacial period and instability during the late-glacial period, only to be followed by restabilization during the early Holocene. Davis (1981) concluded that species populations in temperate regions are inherently unstable and have been in disequilibrium with climate during much of the Quaternary. Delcourt and Delcourt (1983) show continual vegetational disequilibrium during postglacial times north of about 43 degrees north latitude in eastern North America. According to them, throughout the late Quaternary, this vegetation has been in dynamic equilibrium with a relatively constant flora south of 33 degrees north latitude along the Gulf Coastal Plain, peninsular Florida, and west-central Mexico where temperature changes have not been great. Between 33 and 39 degrees north latitude, apparently stable vegetation communties of full-glacial times were replaced by relatively unstable vegetation during late-glacial warming and these were followed by stable communities after 9 kya. We shall see in the following sections just how this pattern of change developed.

At the time of the last major glaciation, much of North America was covered with ice. Vegetation had retreated southward and remained in

refugia—isolated patches of suitable habitat—for extensive periods. As the glacier retreated, species after species moved to the north, with tundra and spruce/jack pine forests bordering the ice. Just as for climates, a proxy record of past vegetation can be found in the sediments of lakes, swamps, bogs, and other depositional sites. Pollen grains from vegetation on the nearby landscape blow onto the surfaces of these bodies of water and then sink to the bottom. Studies have shown that distinctive pollen assemblages correspond with major vegetation types. The assumption is made that the assemblage of fossil pollen grains in a sediment represents the relative abundances of vascular plant taxa in the vegetation community surrounding the depositional site.

Vegetation Migration in North America

Delcourt and Delcourt (1981) drew vegetation maps of eastern North America from 40 kya to the present, as derived from the sediment pollen record (Figures 1–4). Our story begins at 18 kya, the time of the last major extension of the ice sheet. First we must identify the major pollen assemblages associated with each major plant community.

In the modern tundra of northern Canada, herb pollen usually represents more than 25% of the pollen found. Birch, alder, willow, and spruce pollen make up most of the rest. In the boreal (cold northern) forest of eastern Canada today, the percentage of herb pollen is less than it is in the tundra. Boreal forest pollen includes spruce, pine, birch, fir, and larch. The mixed conifer–northern hardwood forest has a distinctive pollen assemblage of hemlock, jack and/or red pine, white pine, spruce, fir, oak, birch, elm, ash, ironwood, maple, and beech. The cool-temperate deciduous forests of the eastern United States today have pollen assemblages dominated by oak and including many other forest taxa, such as hickory, maple, beech, basswood, elm, walnut, hemlock, and gum. There may be some pine present. The pollen assemblages of the warm-temperate southeastern evergreen forest includes oak, hickory, southern pine, and perhaps cypress and gum. Open vegetation communities, identified as prairie on the western border of the maps and as sand dune scrub in southern Florida, are characterized by a high percentage of herb pollen.

18,000 Years Ago

At the peak of the continental glaciation, around 18 kya, sea level was approximately 100 meters lower than it is today. The Wisconsin ice sheet extended across North America from the Canadian Rockies through the northern High Plains and southward across central Illinois and Ohio and into southern New England. A treeless belt of tundra 60

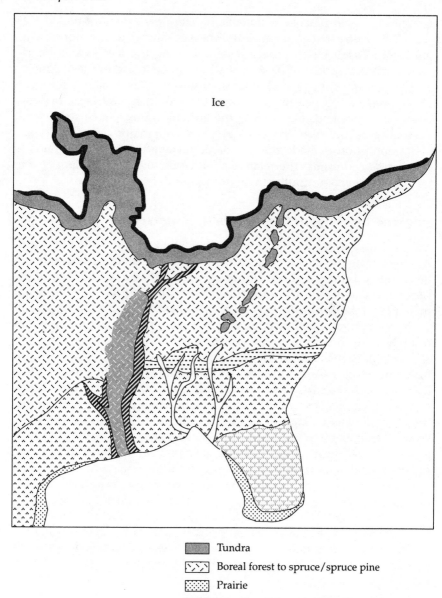

Ice

Tundra

Boreal forest to spruce/spruce pine

Prairie

Mixed conifer-northern hardwood forest

Boreal forest

Cool-temperate deciduous forest

Warm-temperate southeastern evergreen forest

Sand dune scrub

FIGURE 1   **Paleovegetation map for 18 kya. (After Delcourt and Delcourt, 1981.)**

to 100 km wide was adjacent to the ice (Figure 1). The tundra extended in broken fashion down the Appalachian crest to the Great Smoky Mountains. Spruce and jack pine occupied the Great Plains. Jack pine, along with spruce and fir, occupied the interior low plateaus, the middle slopes of the Appalachians, and the Atlantic coastal plain.

A warm-temperate forest of oak, hickory, and southern pine grew on the Gulf and lower Atlantic coastal plains. A northern hardwood and cool-temperate coniferous forest is shown as a narrow transitional zone between jack pine and spruce to the north and oak, hickory, and southern pine to the south. Southern Florida was too dry for anything but prairie and sand dunes. According to Davis (1981), as the climate warmed between 16 and 10 kya, spruce—an aggressive pioneer—moved into tundra while tundra plants continued to stay near the ice edge. In the Great Lakes region spruce became established quickly after the glacier retreated, but in New England there was nearly a 2000-year lag. Spruce appears to have needed nitrogen-fixing plants, such as alder, to precede it into a region.

10,000 Years Ago

By 12.5 kya the climate had warmed sufficiently that the spruce forest within the lower Mississippi alluvial valley was replaced by a cypress and gum forest. The ice sheet had retreated north of the Great Lakes by 10 kya (Figure 2). Tundra still existed along the St. Lawrence Valley and into New England. Spruce forest dominated the Maritime provinces of eastern Canada, New England, and southern Manitoba. Spruce and jack pine extended from Minnesota across the Great Lakes and into southern Quebec. A mixed conifer–northern hardwood forest expanded northward and occupied a broad band from Minnesota to Maryland. There were islands of spruce and fir stranded in the southeastern United States. Open prairie and oak savanna developed in the Great Plains, and oak savanna in Florida.

5000 Years Ago

The world was at its warmest during the hypsithermal interval, from 4 to 8 kya. There was somewhat less solar radiation than earlier, but strong westerly winds prevailed and precipitation became available throughout the growing season in the southeastern United States. Sea level had risen to approximately its modern level. Prairie had expanded eastward and northward, as had oak savanna and oak–hickory (Figure 3). White pine moved northward, and the mixed conifer–northern hardwood forest entered southern Canada. Oak and hickory were replaced by southern pine on the sandy soils of the Gulf and Atlantic coastal plains. Southern pine replaced oak savanna in Florida. About 6 kya or

Tundra

Boreal forest to spruce/spruce–pine

Prairie

Mixed conifer-northern hardwood forest

Cool-temperate deciduous forest

Warm-temperate southeastern evergreen forest

FIGURE 2 **Paleovegetation map for 10 kya. (After Delcourt and Delcourt, 1981.)**

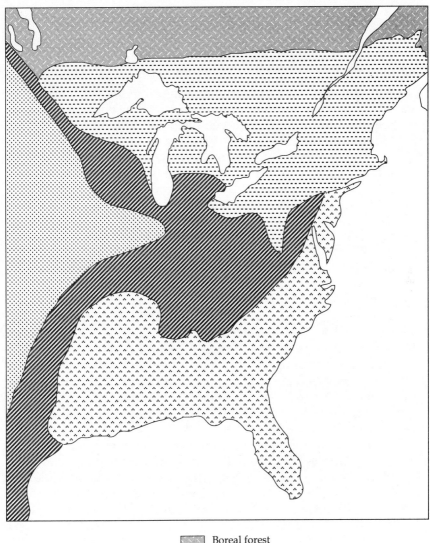

Boreal forest

Mixed conifer-northern hardwood forest

Prairie

Cool-temperate deciduous forest

Warm-temperate southeastern evergreen forest

FIGURE 3   **Paleovegetation map for 5 kya. (After Delcourt and Delcourt, 1981.)**

earlier, vegetation communities were 100 to 300 km farther north than they are today, and mean annual temperatures were 1.5°C or more warmer.

200 Years Ago: Presettlement Vegetation

The climate cooled considerably after the hypsithermal, and by 200 years ago solar insolation was near its modern level. The boreal forest southern boundary had moved southward and the prairie–forest boundary had moved westward (Figure 4). (The boreal forest movement is described in more detail in a later section.) Relic prairie pieces were left to the eastward. Coastal swamps and marshes developed in southern Louisiana as the Mississippi River delta formed. The mixed conifer–northern hardwood forest was invaded by oak and hickory and other hardwoods from the south.

Climate Forcing of Forests

An **air mass** is an entity of the atmosphere that is approximately homogeneous in its horizontal distribution of temperature, humidity, and lapse rate. The natural boundaries between air masses form **frontal zones** or **fronts**. Delcourt and Delcourt (1983) worked out the patterns of biotic responses to the air mass boundaries that, according to Bryson (1966), define climate regions. These generalized associations are shown in Figure 5. The cool temperatures that prevailed at full glaciation resulted in reduced evaporation of water from the oceans and decreased precipitation over the continent due to the diminished frequency of tropical depressions and hurricanes. The boreal forest was under the influence of Pacific air masses while the ice sheet was to the north. About 14 kya the transition zone between the boreal and temperate forests had become established and was associated with the mean winter position of the polar front. After the retreat of the ice sheet, around 11 kya, the northern and southern boundaries of the boreal forest coincided with the mean summer and winter positions of the Arctic air masses. The cool-temperate deciduous forest was associated with maritime tropical air masses and Pacific air masses. The warm-temperate southeastern evergreen forest was under the influence of maritime tropical air masses.

Rates of Forest Migration in North America

The change in the types of species and numbers of individuals of a given species in an ecological community over time is referred to as **succession**. An "individualistic" view of plant communities suggests that succession results from the interaction of individuals and species responding to local conditions of soil and climate. By contrast, the

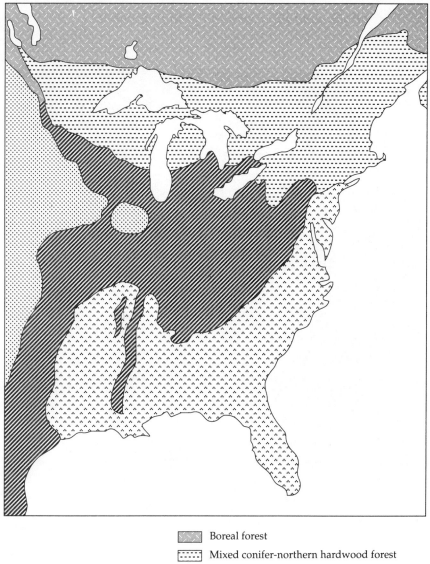

	Boreal forest
	Mixed conifer-northern hardwood forest
	Prairie
	Cool-temperate deciduous forest
	Warm-temperate southeastern evergreen forest

FIGURE 4 **Paleovegetation map for 200 years ago. (After Delcourt and Delcourt, 1983.)**

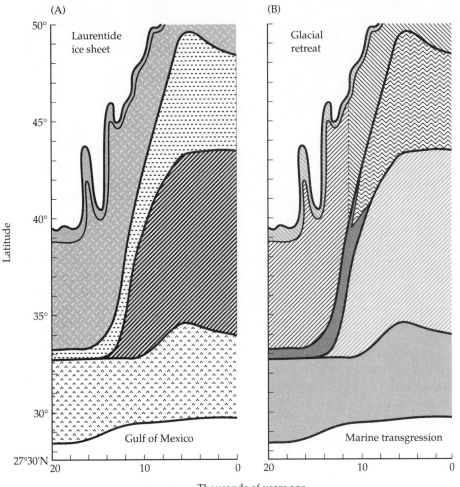

Tundra

Boreal forest

Mixed conifer–northern hardwoods forest

Cool-temperate deciduous forest

Warm-temperate southeastern evergreen forest

Ice margin chill zone

Pacific air mass

Arctic air mass

Arctic and Pacific air mass

Maritime tropical and Pacific air mass

Polar frontal zone

Maritime tropical air mass

◄ FIGURE 5 Schematic diagram showing changes in (A) vegetational patterns and (B) environmental forcing functions as air mass boundaries defining climate regions on a transect along 85° west longitude in eastern North America over the last 20,000 years. The polar frontal zone shown is for the mean position of the polar front in winter. The vertical dashed line at 11,000 years in (B) represents the removal of the ice sheet as a significant topographic barrier to the southward flow of arctic air. (After Delcourt and Delcourt, 1983.)

"Clementsian" view (Clements, 1916) suggests that the entire biotic community develops like a single "superorganism," but with communities of plant species on a given site changing over time as they compete with one another and evolve to a climax stage. (See Odum, 1989, for a full discussion of plant succession.)

Davis (1981) writes:

The history of the spread of trees northward during the present interglacial leads inevitably to an individualistic view of plant communities. Even forests mapped as a single community have had very different histories. For example, oak–chestnut forests in the central Appalachians have included chestnut as a dominant for 5000 years or longer, while oak-chestnut forests in Connecticut have included chestnut for only 2000 years. Deciduous forests in Ohio were penetrated first by hickory and then by beech 4000 years later; in Connecticut beech arrived first followed 3000 years later by hickory.

In another example, hemlock spread more quickly than beech and arrived in upper Michigan 2500 years before beech did. So before we jump to the conclusion that climate changes are the only cause of forest migration, it is important to consider the conclusion reached by Davis, who states:

. . . much of the time the rate of spread was not controlled by climate, and the geographic distributions of many species were not in equilibrium with climate, depending instead on the availability of propagules and the ability of seedlings to survive in competition with plants already growing on the site.

As the glacier retreated northward, tree species did not all simply follow in that direction. Some moved from east to west, while others moved from west to east, and a few moved northward. Several species moved without any apparent climate limitation in one part of their range, but seem to have been restricted by climate in another region. At its western boundary in Minnesota, white pine was in equilibrium with the climate for several thousand years, but continued to move northward. During much of the Holocene many genera of forest trees were in disequilibrium with climate. Hickory, for example, reached New England well after the warmest part of the Holocene.

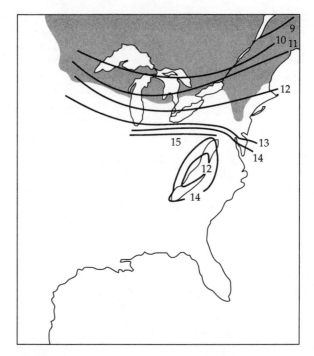

FIGURE 6   **Migration map for spruce (*Picea* spp.). The numbers refer to the age (in kya) when spruce first arrived at a site. The shaded area represents the modern range for spruce. (After Davis, 1981.)**

Migration maps for spruce, white pine, and oak are shown in Figures 6, 7, and 8. Spruce, a boreal species, moved from south to north into the retreating tundra about 2000 years after the ice had left the Great Lakes region. Spruce forests may have been very open and patchy in their distribution in the Middle West and the Appalachians. Oak spread very rapidly between 11 and 9 kya and reached its present range limit about 7 kya. White pine appears to have had a refuge area in the southeastern United States. It reached the Great Lakes region around 10 kya and Minnesota about 7 kya, but then its expansion to the west was prevented by the dry prairie climate. Lack of competition from deciduous tree species in many parts of its range may have allowed white pine to have temporary population increases in places where it is no longer abundant. Both oak and white pine have similar northern limits, but white pine penetrates more into the boreal forest.

Table 1 shows the average rates of Holocene range extensions for 12 genera of trees in eastern North America. The rates at which trees migrated in response to climate are much slower than the rate of climate

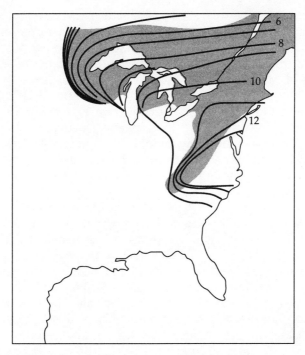

FIGURE 7 **Migration map for white pine (*Pinus strobus*). The numbers refer to the age (in kya) when white pine first arrived at a site. The shaded area represents the modern range for white pine. (After Davis, 1981.)**

TABLE 1 **Average rates of Holocene range extensions in eastern North America.**

Species	Rate (m/yr)
Jack/red pine	400
White pine	300–350
Oak	350
Spruce	250
Larch	250
Elm	250
Hemlock	200–250
Hickory	200–250
Balsam fir	200
Maple	200
Beech	200
Chestnut	100

(From Davis, 1981.)

change. It may be that soils need a certain degree of development before they can support a tree species. The table shows that tree genera move at rates of 10 to 40 km per century. The southern border of spruce took from 10 kya to 5 kya to move from the southern border of Michigan to the Straits of Mackinac, a distance of 500 km. This is a migration rate of 10 km per century. By contrast, the mean annual isotherm could move poleward at 85 km per °C of temperature, or about 250 km in the next century if we assume a 3°C temperature increase within the next 100 years due to greenhouse warming. This rate of isotherm movement is an order of magnitude faster than the rate of range extension for forest trees in the past and suggests that future forests may not keep pace with the expected change.

Gear and Huntley (1991) describe the presence of Scots pine in northern Scotland 4400 years ago for a period of about 400 years. At about that time there was a change in atmospheric circulation with a northward shift of the jet stream and of the Azores high pressure system, resulting in decreased rainfall and warmer summer temperatures in northern Britain and Scandinavia. Then about 4000 years ago the pines began to die, and by 3800 years ago the pine forests had retreated to their earlier northern limit, where they are found today. This was a migration of about 80 km northward, then southward at a rate between 37.5 and 80 km per century. This rate is comparable to those Davis described, and is also too slow to keep pace with the current rate of greenhouse warming.

Propagules

Forest trees reproduce from seeds and/or root sprouts. If reproduction is by root sprouts, as with aspen, then the rate of migration is expected to be very slow. If the seeds are blown by the wind or transported by animals, the rate of movement may be much faster. However, experiments with seed dispersal show considerably slower rates of movement than even the Holocene pollen record indicates. Measurements show that only 5% of spruce seeds are carried as much as 200 meters downwind from the parent tree. This would represent incredibly slow dispersal if a new tree had to grow to maturity in order to extend this distance. Seeds of the grass *Agrostis hiemalis* have been observed to move 9 meters per year—less than 1 km per century.

A climate beyond the range of a particular species may be suitable for its growth, and even for its reproduction, but unless the species can move into that region sufficiently rapidly, it may be left behind as suitable climate moves beyond its reach. Dispersal is likely to be even more of a problem in the modern world than it was in the past as many

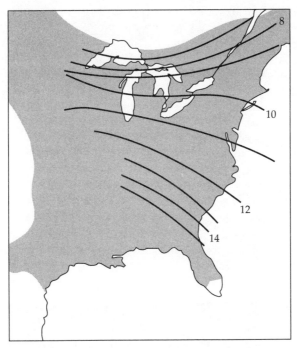

FIGURE 8   Migration map for oak. The numbers refer to the age (in kya) when oak first arrived at a site. The shaded represents the modern range for oak. (After Davis, 1981.)

barriers, such as cities, highways, and agricultural fields, stand in the way of plant migration.

When beech and hemlock moved into the Great Lakes region of North America during the Holocene, the lakes were a major barrier to their migration (50 to 100 km wide). According to Davis (1987), the fossil pollen record shows that beech arrived at the western shore of Lake Michigan, either by long-distance dispersal across the lake or by moving from one habitat to another across the prairie landscape south of the lake. Whatever happened is not known, but it took 500 to 1000 years from its establishment on the eastern shore to reach the western shore of Lake Michigan. Beech nuts are dispersed by birds and mammals. Blue jays feed on beech nuts and often cache them in forest leaf litter, where they may germinate and grow if the birds fail to dig them up and eat them later. Blue jays are known to transport beech nuts to their nesting areas over distances as great as 4 km. However, since beech trees take at least 50 years to mature and bear fruit, it is hard to see how they could move at 20 km per century. More likely, a bird such as the passenger pigeon, which is now extinct, may have provided the long-distance flight necessary to get the beech nuts across Lake Michigan in one jump and the numbers to repeat the journey enough times so that success was assured.

Hemlock dispersal is by wind-borne winged seeds that fall from the tree during the autumn and winter months. The seeds may then be blown across a frozen lake or snow surface for considerable distances, such as across Lake Michigan; Davis (1987) suggests distances of tens of kilometers. Hemlock seedlings best take root in rotting logs or stumps and other moist habitats. Hemlock was established in southern Ontario 8 kya and appeared in northern Michigan 7 kya. Its migration route into the Straits of Mackinac region was from the east, not from the south. The western limit of hemlock was established by 2.5 kya and was stable for the next 1500 years; then during the last 1000 years the species expanded its range, particularly as the Little Ice Age climate became favorable to it.

Vegetation Responses to the Younger Dryas Event

The Younger Dryas event was a period of dramatic climate change, as described in Chapter 2. At its inception, about 11 kya, a 6°C cooling set in over a 100-year period in the North Atlantic and adjacent land areas. At the end of the Younger Dryas cold, about 10.72 kya, a warming of as much as 7°C occurred in 50 years. These rates of temperature change are comparable with, or actually more rapid than, the temperature change expected over the coming century from global warming. Atkinson et al. (1987) showed that the intense cold of the Younger Dryas was followed by a warming of 7°C to 8°C in summer and 25°C in winter in Britain.

In northwestern Europe, trees (birch and spruce) that had started growing in response to the climate warming of the early Holocene were suddenly replaced by shrubs and herbs during the Younger Dryas period. Tundra returned to much of Europe from northern England to central Germany, although some trees remained in refugia in Denmark and pines persisted in south Germany.

The Younger Dryas in Britain

Atkinson et al. (1987) reconstructed the climate of Britain over the last 22,000 years using remains of 350 beetle species found in ancient sediments. They assumed that if they knew the present-day climate tolerance range of each beetle species, then they could infer the paleoclimate from fossil assemblages containing the same species. Coope (1977) has

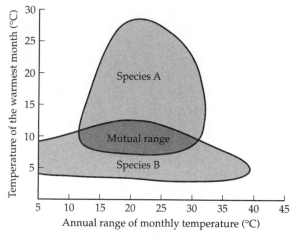

FIGURE 9 Tolerance ranges and mutual climate range of two species of beetles (Coleoptera). (After Atkinson et al., 1987.)

shown that individual beetle species respond to the thermal climate of their environment, especially to summer temperatures and the degree of seasonality between summer and winter. Thus if several species occur together as fossils, the paleoclimate in which they were living can be defined by the mutual intersection of their individual tolerance ranges (Figure 9). Using this technique, Atkinson and colleagues reconstructed the mean temperatures for the warmest and coldest months of the year. They used carbon 14 to give them the dates for the fossil remains. The results are truly striking for their detail (Figure 10). Extremely rapid warmings show up in their results at 13 and 10 kya and coolings appear from 12.5 to 10.5 kya.

The ice sheet had vanished from western England and Wales by 14.5 kya and from Scotland by 13 kya. During this 1500-year period it was very cold in Britain. The coldest winter months were −20°C to −25°C. The climate had more seasonality, that is, was more continental, than it is today. The winter-to-summer temperature range was 30°C to 35°C, compared with 14°C at present. Such extremes are found today in northeastern European Russia and northwest Siberia. Around 13.3 kya, the polar front was moving, over hundreds of years, from its glacial position off the coast of Portugal to a position between Greenland and Iceland. By 12.5 kya the climate of England and Wales was as warm as the present-day climate, although somewhat more continental, with the warmest month up to 17°C and the coldest month about 0°C. Cooling set in from 12.5 to 12.0 kya and was followed by 500 years of fairly constant summer temperatures. But winter temperatures in the coldest

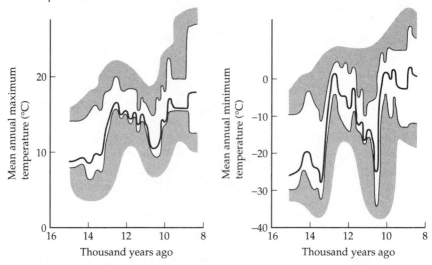

FIGURE 10   **Mean annual maximum and minimum paleotemperatures in Britain inferred from radiocarbon-dated beetle remains. The bold lines indicate the most probable temperature values. The shaded areas indicate the mutual climate ranges (see Figure 9) used to determine the temperatures. (After Atkinson et al., 1987.)**

month decreased, from −5°C at 12.0 kya to −17°C at 10.5 kya. Summer temperatures dropped after 11.4, kya reaching a minimum monthly mean of 10°C about 10.5 kya. This probably corresponds to Dansgard's estimate of 10.7 kya as the end of the Younger Dryas (see Chapter 2). By 9.8 kya the climate was as warm as it had been about 13.3 kya.

The Younger Dryas in North America

According to Davis (1981), spruce arrived in the Great Lakes region soon after the retreat of the glacier, but in New England, where deglaciation was rapid, it took spruce nearly 2000 years to establish itself. Its first appearance in New England was about 12 kya. At Alpine Swamp, New Jersey, oak, ash, and hornbeam arrived at about the same time, while spruce and pine were there much earlier, according to Peteet et al. (1990). Around 11 kya there was a strong decrease in the abundance of the hardwood species and an increase in spruce, fir, larch, paper birch, and alder. About 10 kya oak began to increase in the pollen record. There is only a slight suggestion of the Younger Dryas event of cooling and sudden warming in this vegetation response.

Across the North American continent to the west, at Glacier Bay in southeastern Alaska, an early Holocene pine parkland was replaced about 10.8 kya by a shrub/herb dominated tundra, which lasted until about 9.8 kya, according to Engstrom et al. (1990). Although this suggests possible Younger Dryas influence there, they write:

A distinctive stratigraphic reversal in the pollen percentages of spruce and other northern trees in Ohio indicates that the climatic effects extended upwind from the North Atlantic at least 100 to 1000 km. In contrast, a Younger Dryas event is not apparent in the contemporaneous spruce forest between Nova Scotia and Ohio, presumably because vegetational ecotones were absent from the region at the time.

The Younger Dryas cold event has been particularly informative concerning the response of vegetation to the somewhat slower cooling in the beginning and the abrupt warming at its termination. The cooling appeared to affect forest trees quickly, persumably as an effect of freezing, as there was very little lag in response. When conditions warmed quickly the forest trees were much slower to advance again.

Plant–Climate Response Surfaces

The pollen content of sediments reflects the vegetation communities of the past. If in addition one can determine the response of each plant species to climate variables, one then has a proxy measure of the climates extant when the sediments were deposited. Vegetation on a large scale has considerable inertia over time (changes very slowly) and therefore can integrate climate into long-term averages of several decades to as much as a century. Bartlein et al. (1986) mapped the modern-day distributions of eight pollen types in the eastern North American vegetation and overlaid these with maps of mean July temperature and annual precipitation distributions. They then formed three-dimensional surfaces by means of polynomial regressions of pollen frequencies (percentages) against the two climate variables to obtain response functions for each taxon. Webb et al. (1987) added a third climate variable, the mean January temperature. Usually the pollen record cannot distinguish individual species within a genus; it will show all the oaks together, or the pines together, for example. However, when Bartlein et al. formulated the transfer functions for pine, they obtained a distinct bimodal distribution with a peak response at 17°C mean July temperature for the pine pollen of the mixed conifer–northern hardwood forest and 28°C for the pine pollen in the southeastern evergreen forest.[*]

Figures 11A and 12A show the response surfaces for spruce pollen. It has a single optimum at annual precipitation values of 800 to 1200

[*]Much of this work is based on Thompson Webb's many years of pollen sampling and analysis across a vast region of eastern North America. See Bernabo and Webb (1977) and Webb (1987).

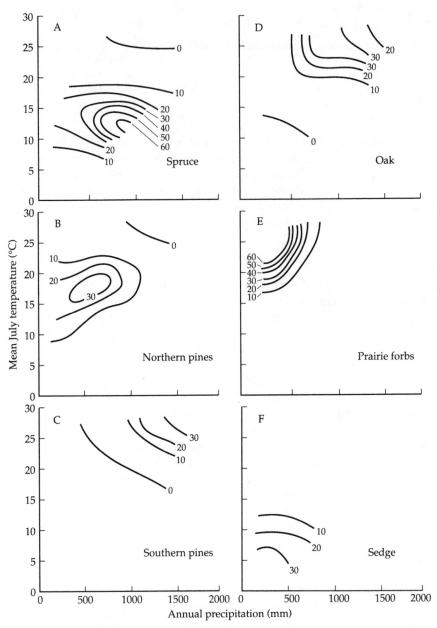

FIGURE 11 **Response surfaces showing the relationship between the percentages of six pollen types as a function of mean July temperature and annual precipitation. (After Webb et al., 1987.)**

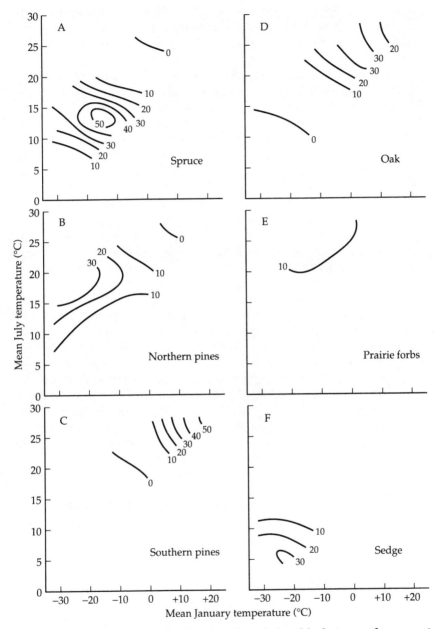

FIGURE 12 Response surfaces showing the relationship between the percentages of six pollen types as a function of mean July temperature and mean January temperature. (After Webb et al., 1987.)

mm and mean July temperatures of 10°C to 13°C—the present climate conditions of eastern Canada. Spruce abundance becomes less with increasing temperature southward and with decreasing precipitation northward.

As mentioned earlier, pine pollen exhibits two optima, since no distinction was made between northern and southern pines. The optimum conditions for northern pines, seen in Figures 11B and 12B, require 300 to 700 mm of annual precipitation and mean July temperatures of 15°C to 20°C, corresponding in modern times to a region from western Ontario across the northern United States into New England. The optimum for the southern pines, seen in Figures 11C and 12C, required 1250 to 1700 mm of annual precipitation and mean July temperatures of 25°C to 30°C, characteristic of the southeastern United States. The contours around this optimum show a more rapid decrease in pollen frequency with decreasing temperatures than with decreasing precipitation. Where temperatures were high, the abundance of the southern pine diminished as annual precipitation dropped below 1000 mm.

Oaks have the response surfaces shown in Figures 11D and 12D. The optimum is at mean July temperatures 23°C to 27°C and mean annual precipitation of 750 to 1250 mm. The oaks clearly prefer warm summers and moderate amounts of annual precipitation. These conditions are to be found from the Ozark Mountains and the middle Mississippi Valley eastward. The oak genus represents many species, which readily hybridize. It would be interesting to see response functions for individual species. Figure 12D shows that some oaks survive at mean January temperatures below −10°C, but only in low abundance. The ranges for the red oak and the bur oak extend into the upper peninsula of Michigan, where temperatures are quite cool, while pin oak, chinquapin oak, and black oak have northern limits in the lower half of the southern peninsula of Michigan. Some oak species extend all the way south to the Gulf of Mexico.

Prairie forbs, characteristic of the nothern cool-temperate prairie, include sage, various composites, and pigweeds. The response surfaces for prairie forbs are shown in Figures 11E and 12E. These are dramatically different than those for the forest tree genera. Clearly the forbs require dry conditions and cool to warm summer temperatures. The prairie forbs will survive at low mean annual January temperatures, but in low abundance. By contrast, sedge (Figures 11F and 12F) requires low mean July temperatures, low annual precipitation and very low mean January temperatures.

The response surface for birch is shown in Figure 13. One optimum for birch is located at mean July temperatures around 15°C to 17°C and annual precipitation of 1000 to 1500 mm. This represents the paper

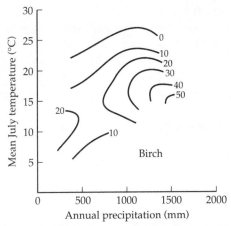

FIGURE 13 Response surface for birch (*Betula*) as a function of mean July temperature and annual precipitation. (After Bartlein et al., 1986.)

birch and yellow birch of the mixed conifer–northern hardwood forests. These species are known to be especially sensitive to climate warming and drying. They are readily stressed under these conditions and become vulnerable to infestations by the birch borer, which results in their death. The other optimum seen in the response surface is at a mean July temperature of 10°C and low precipitation, about 300 mm. These conditions represent the dwarf birch of eastern Canada and the Arctic.

Using these response surfaces and mathematical transfer functions, along with pollen sediment records, Webb (1987), Webb et al. (1987), and Prentice et al. (1992a) reconstructed the vegetation and climate conditions in eastern North America during the past 18,000 years. Differences between the model and observational results suggest caution in using the model to interpret the climate change during the last deglaciation.

Boreal Forest Movement

The boreal forest and its **ecotones**—areas of transition between different types of habitat—are of particular importance in our study of climate change and ecosystems. The greatest greenhouse warming in the future is expected to be at high latitudes. The boreal forests will be affected strongly by this change. The effects of past climate changes on the boreal forest may give us some insight into the future changes to be expected.

In North America, the boreal forest has a northern ecotone with the

tundra and a southern ecotone with the mixed conifer–northern hardwood forest in the eastern part and with the prairie in the western part
of its range (see Figure 4). The boreal forest covers a well-defined climate
region described by Bryson and Murray (1977) as follows:

The spruce forest must have arctic air in winter. In summer it must have Pacific
air, dry air coming down from the western mountains. Farther north, under
arctic air year round, lies the tundra. To the south of the spruce, where moist
tropical air reaches in summer, are forests of mixed hardwoods. Or, where dry
air from the west prevails, lie grasslands.

The air flowing over upper North America originates to the west
and north during the winter and may, during the summer, also originate
to the south and somewhat to the east. The flow of the westerlies
contains the jet stream, which meanders in great loops or waves about
the pole. In winter, the westerlies are in their most southward position
and arctic air dominates a large region. In summer, the westerlies are
more contracted, arctic air is confined to the north, and warm tropical
air can push up from the south. The form of the loops varies from year
to year, from month to month, and with major climate shifts. A certain
region can be under a flow from the west, north, or south, depending
on the arrangement of the waves. Bryson (1966), in a beautiful paper
analyzing the flow of air over the Canadian boreal forest, wrote:

That fronts and wave cyclones are more common in the general vicinity of the
treeline than over either the tundra or boreal forest proper is rather easy to
demonstrate. . . . This implies that the dominance of a particular air mass over
the tundra and a different air mass over the boreal forest may be of more than
casual interest.

Bryson produced an elegant analysis of the air masses and fronts over
the Canadian tundra and boreal forest. He followed air mass trajectories
to the borders of Canada and defined the masses according to their
origin as Arctic, Pacific, Atlantic, and United States. He then obtained
the frequency with which each air mass is found over various parts of
Canada for each month of the year. Air originating over the Arctic
Ocean occurs with the highest frequency over the Canadian Arctic
Archipelago in the far north. Pacific airflow has a close association with
the prevailing westerlies and occurs most frequently west of the Canadian Cordillera. It decreases eastward, is strongest in the Prairie
Provinces and weakest east of Hudson Bay. The westerlies carry Pacific
air masses deep into the continent. Churchill, Manitoba, located more
than 700 km from the region of origin, has Pacific air one-third of the
time, which is almost equal to the frequency of Arctic air and much
more frequent than Atlantic air. Air of Atlantic Ocean origin does not
penetrate very far to the west on the continent. The eastern shore of

Hudson Bay receives Atlantic air only 20 percent of the time. Air having its origin over or beyond the continental United States enters Canada in the Great Lakes and Great Plains areas. South of Hudson Bay the air originates over the United States one-third of the time in July.

Bryson found that the boreal forest lies between the summer and winter positions of the Arctic front (Figure 14). The Arctic front is anchored the year round at the northern end of the Canadian Rockies, but swings north and south with the seasons over the flat continental

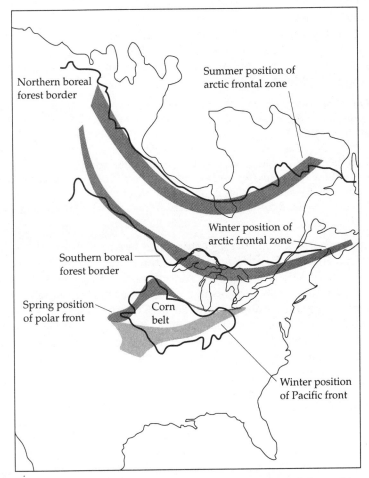

FIGURE 14 The coincidence of the North American boreal forest biome with meteorologically defined climate (air mass) regions. The corn belt of the mid-western United States (see Chapter 7) is also shown as a coincidence of air masses. (After Bryson, 1966.)

interior to the eastward. Bryson defined the Arctic front as the dynamic boundary between the cold Arctic (and continental Polar) air masses and the Pacific air masses. Arris and Eagleson (1989), and earlier George et al. (1974), observed that the southern ecotone of the boreal forest approximately coincides with the $-40°C$ average annual minimum temperature isotherm. Freezing temperatures play a major role in limiting some plant distributions. A few plants are able to withstand very cold temperatures, as low as $-40°C$. The $-40°C$ isotherm is the northern limit for the mixed conifer-northern hardwood forest. The boreal forest is adapted to much colder temperatures than $-40°C$; boreal forest seems unable to compete with the tree species of the mixed conifer-hardwood forests south of this line. Curiously, this apparent relationship does not seem to hold in Europe or Asia, where the boreal forest grows well south of the $-40°C$ average annual minimum temperature isotherm.

Bryson and Murray (1977) wrote:

North of the present forest, in a zone stretching 1000 miles from west to east, are many areas where layers of podzolic soils alternate with the grey-brown soils of the treeless tundra. In these soils are written the comings and goings of the forest as the arctic air environment expands and contracts. A thousand years ago the forest had advanced northward into the barren lands, the tundra. Then something happened to reverse this advance. The arctic was expanding again, and for the area that lies generally north of the Great Plains, this last major expansion is recorded in the soils from about the beginning of the thirteenth century.

The northern boreal forest ecotone has been north or south of its present position in Canada at different times in the past and has migrated as the Arctic front moved. According to Ball (1986) there was a shift in the pattern of winds and in the number of days of rainfall and thunder before and after about 1760 at Churchill and York Factory, Manitoba. This is a clear indication of a change in the mean summer position of the Arctic front; a change resulting from a shift in the general circulation. The Little Ice Age was slowly coming to an end about then.

In the region of Churchill the Arctic front, and therefore the tree line as well, moved to a more northerly position during the nineteenth century. According to Ball, Samuel Hearne drew maps for the Hudson's Bay Company showing the location of the tree limit in 1772. He mapped the tree line all the way from north of Churchill to above Great Slave Lake. This map shows the tree line far south of where it is today—by as much as 120 to 220 km, depending on the location. Ball described the meteorological conditions existing at Hearne's time showing how the Atlantic surface trough had a more westerly position than it does today.

Ball raised three questions. First, how closely does the tree line approximate the climate conditions that appear to be its cause? The answer is that it does so very well. Second, how quickly would the tree line respond to a change in those climate factors? The answer to this question is, slowly and with a lag. Various investigators, including Bryson, found tree outliers, or "tree islands," beyond the present-day tree limit, which seem to be relics of warmer conditions before the Little Ice Age. Along the tree line, although there would be some years when climate conditions would be favorable to seed production, the longer-term conditions might be unfavorable and therefore the forest would only respond vegetatively. The climate change would have to be of sufficient magnitude and length to allow production of seed and its survival. The third question is, how dramatic would the change in climate have to be to bring about a shift in the tree line? Nichols (1975) indicates that a change of the mean July temperature of +4°C or −1°C may produce northward or southward movement of the boreal treeline. Kay (1979) found that the boreal forest extended as far as 62° north latitude during the hypsithermal period, when the mean July temperatures were 1°C to 3°C higher than they are today, after which there was a southward movement of the tree line in the vicinity of Long Lake, north of Churchill.

Rates for Individual Species

It is useful to study the changes of tree species distributions on mountains, since one can move to colder or warmer climates by simply going up or down the mountain. Kullman (1979) did this when studying the distribution and establishment of birch (*Betula pubescens*) in central Sweden. During a period of climate warming from 1925 to 1950, the rate of birch establishment was relatively slow, with a gradual increase over a 20-year period as birch moved up the mountains into high altitude tundra. When the climate began cooling after that, the rate of decrease for birch at the higher altitudes was rapid, with a significant decrease occurring in 5 years. This suggests that birch would move progressively, and noticeably, into tundra during climatic warming. Ten to 20 years of observation might detect a change.

In contrast to the study of birch along its ecotone with tundra, Hamburg and Cogbill (1988) evaluated the composition of old-growth stands in the spruce–hardwood forests of central New Hampshire and showed changes within the stands. Their data show a major decrease in canopy dominance by red spruce from 1800 to the present and an increase in the presence of beech. They suggested the hypothesis that the major driving force for this change has been a warming trend in both the mean annual and mean summer temperatures. This may have resulted in increased competitiveness of beech relative to spruce.

Lessons for the Future

North American populations of forest trees north of 43° north latitude are inherently unstable and have been in disequilibrium with climate during much of the Holocene. Forests move slowly in response to climate change; rates of individual species movement have ranged from 10 to 40 km per century. Reproduction and seed dispersal is a limiting factor. Cooling of the climate may affect forests quickly through freezing and shortening of the growing season, but during periods of warming forests are slow to move. Some trees may reproduce vegetatively and remain in a region for a long time after adverse conditions prevail for reproduction from seed.

Boreal forests are especially good subjects for study since in North America their northern and southern ecotones are located along the summer and winter positions of the Arctic front. In the past climate change has produced movements of the Arctic front, and the boreal forest has moved with it, but with a time lag. If GCM models show how the frontal systems will move with greenhouse warming, then one can anticipate the changes which may occur along the forest ecotones.

The pollen record has been useful for deriving response surfaces for various tree species as a function of temperature and moisture availability. These response surfaces show the optimum conditions for selected species or genera. These will be useful when discussing the impact of greenhouse warming on forest trees, grasslands, and sedges. All of the above will be incorporated in the discussion of forest gap models in the next chapter.

Forest Models
and the Future

FORESTS COVER between 35% and 40% of the Earth's land area, produce about 65% of the carbon fixed annually, and store over 80% of the world's organic carbon. The response of forests to increasing levels of carbon dioxide and to increasing temperature and changing precipitation patterns is of vital importance to the future well-being of the world's people. Natural forests of mixed species and mixed age make up the major part of timber resources; intensively managed tree plantations play only a small part. Although plantation plantings are more intensively managed in Europe and Japan, the total area of the world's forests created by planting was about 3% of all closed forest areas in the late 1970s. Tropical forests occupy about 13% of the Earth's land surface, but contain more than 90% of the world's species diversity.

The Europeans have a long tradition of forest management; recently forest decline throughout much of Europe has been of enormous concern and has led to intensive studies of air pollution and forest health. Forests are our only source of wood, our major source of paper, and the origin of many chemicals. Forests contribute habitat for wildlife, protect watersheds against erosion, and provide wilderness areas for our enjoyment that also act as buffer zones against human crowding.

Trees are long-lived and take a long time to reproduce. It is not easy for them to respond quickly to environmental changes, nor is it easy for us to develop trees adapted to changed conditions through genetic selection. Natural forests may lag in their response to changing conditions. Reproduction may fail as the climate warms, or as it gets wetter or drier, but this may not be evident for many decades. Increased storminess—more hurricanes, tornados, and downbursts—will have an enormous impact on forests. Disturbance has always been a part of natural forest dynamics, but too much disturbance is extremely destructive. The competitive balance among forest tree species will change as some species are better adapted to the changing conditions and others less so. Forest tree species migrate only slowly across the landscape as climate changes, but in the future their routes of migration will be interrupted by highways, cities, and agricultural fields.

In order to address the many questions related to the future of forests, it is necessary to model their responses to changing climate and environmental conditions. We cannot wait to see what will happen, nor can we expect to manage our forests against all future exigencies until we have a more basic understanding of what may happen. (See Gates, 1990 for a brief discussion of climate change and forests.) Our discussion in this chapter is concerned with the natural, mixed-species forests that occupy 97% of the world's closed-canopy forest resources rather than uniform age, single-species tree plantations.

Forest Modeling

Two general approaches characterize the mathematical models that have been used to describe how forest trees respond to climate change. The first approach uses basic knowledge of species' natural history and physiological processes to simulate the responses of species and ecosystems to climate change. These simulation models are also known as **forest gap models**. The second is an empirical approach that correlates the geographic distributions of climate variables with those of ecosystems. This approach is pragmatic and straightforward, but does not depend on a fundamental understanding of the underlying processes. Both approaches require the estimation of future climate scenarios in order to drive the models. The simulation approach will be described in the following pages. Later in the chapter, various empirical models will be described.

The JABOWA Model

The first successful simulation of the population dynamics of the trees of a mixed-species, mixed-aged forest from a conceptual basis was accomplished by D. B. Botkin and colleagues (Botkin et al., 1970; Botkin, 1972). This model, known as the Botkin-Janak-Wallis simulator and given the acronym JABOWA, is a stochastic computer model that reproduces the random characteristics of a forest. It was originally applied to the trees growing in the Hubbard Brook Experimental Forest, located in the White Mountains of northern New Hampshire. Other forest models have been developed since that time, but all of them have started with the basic features of JABOWA.

Tree species are entered into the model using the following characteristics:

1. maximum age
2. maximum diameter
3. maximum height
4. a relation between height and diameter
5. a relation between total leaf weight and diameter
6. a relation between rate of photosynthesis and available light
7. a relation between relative growth and a measure of climate
8. a range of soil moisture conditions within which the species can grow
9. the number of saplings that enter the stand under shaded or open conditions

The abiotic environment is defined by:

1. elevation
2. soil depth
3. soil moisture-holding capacity
4. percentage rock in the soil
5. a set of average monthly temperature and precipitation records calculated from a nearby weather station
6. a value for the annual insolation above the forest canopy

For each year, the model added growth to some trees, added saplings, and killed off other trees. Direct competition among individuals was restricted to competition for light.

The subroutine GROW contained an equation for optimal tree growth (defined as the volumetric change, or the annual change in diameter squared times height) that was directly proportional to the tree's leaf area and inversely proportional to the amount of nonphotosynthetic (nonleafy) tissue. As diameter increased, the rates of change of diameter and height decreased. The change in diameter used was proportional to current diameter and height and the particular species maximum diameter and height; the shading leaf area, which took into account whether the species was shade-tolerant or shade-intolerant; the photosynthesic rate as a function of growing degree days (GDD); the total trunk basal area on the plot, and the maximum basal area under optimum growing conditions. A parabolic function was used for the temperature response of photosynthesis. The points where the parabola crossed zero were derived from the present geographic range of the species and the average monthly temperatures at these limits.

The subroutine BIRTH selected a random number of saplings below a species-specific maximum for each species when shade, GDD, and soil moisture allowed growth of that species. The subroutine KILL used the assumption that no more than 2% of the saplings of any one species would reach their maximum age and that death probability is a linear function of a tree's age and maximum life span. This subroutine took into account random occurrences external to the tree, such as storm damage and environmental factors that might kill off a tree. Also, a tree with a growth rate below a minimum threshold had a 36.5% chance of death in a particular year. The subroutine SITE described the particular site characteristics for the area of interest. It calculated GDD from the closest National Weather Service data base and adjusted for differences in elevation using a January average lapse rate and a July lapse rate. Evapotranspiration rate, the amount of surface area covered with rock, and monthly precipitation amounts were additional site characteristics.

The model successfully reproduced forest succession, competition, and changes in species composition for the Hubbard Brook forest using 13 species. The model predicted a general increase in tree density, or stems per unit of area, and a decrease in basal area with elevation. The model predicted correctly the change from hardwoods to conifers in a realistic elevational range. The growth behavior of individual species, the average basal area per plot as a function of time, was computed for elevations ranging from 458 to 1067 meters. The calculated results agreed remarkably well with the observed forest characteristics. The model also predicted a composition and density for old-age stands that reasonably resembled what was known about some of those stands prior to cutting at Hubbard Brook. However, there were some serious discrepancies, partly caused by the considerable inhomogeneity of the terrain and its history. Botkin (1981) refined the JABOWA model over the years and introduced soil nitrogen as a site characteristic. He stated that the model ". . .appears to be realistic in that it qualitatively reproduces all the major characteristics known for forest succession."

The FORET Model

Shugart and West (1977) made several modifications to the JABOWA model in order to model a forest stand in Tennessee. The changes made in this model, FORET (Forests of East Tennessee), were a nearly three-fold increase in the number of tree species (from 13 to 33); a change in plot size and shape from JABOWA's 10 m × 10 m square to a one-twelfth hectare circle; the removal of some subroutines, including SITE; and the addition of two new subroutines, SPROUT and GAUSS.

The GROW subroutine in FORET was identical to the one in JABOWA. The FORET model did not include soil moisture and therefore was limited to lower slope sites with adequate moisture. On the other hand, it did include GDD variance (using random GDD values drawn from a normal distribution with mean and standard deviation defined by the nearest weather station history). Hence, GROW was a stochastic model, rather than a deterministic model like JABOWA. The BIRTH subroutine selected the availability of mineral soil and the presence of leaf litter, determined by the calculated leaf area index; a positive deviation of temperature from the average year, determined from GDD; and the probability of deer or small mammal populations, determined randomly. Based on these factors, species and numbers of saplings were chosen randomly and added to the stand. Some early-successional species were eventually limited by lack of seed sources and were no longer considered if they had missed being "planted" for 20 or 30 years. The subroutine SPROUT used vegetative reproduction to add trees to a stand if a tree of the proper size and species was killed off by the KILL subroutine. The subroutine GAUSS randomly selected a value of GDD from the modal distribution representing the particular site.

Solomon et al. (1981) used the FORET model to simulate forest succession in response to long-term climate change over the past 16,000 years. They tested the model against the pollen record from Anderson Pond in White County, Tennessee. They increased the number of tree species to 65. The simulation was run under the following four conditions: (1) seed sources from all species were available all the time with no long-term climate change; (2) seed sources were available according to a time sequence with no long-term climate change; (3) seed sources from all species were available all the time along with a climate change; and (4) seed sources were available according to a time sequence along with a climate change. For the conditions in which the seed sources became available over time, seeds were introduced into the model according to the date when each species first appeared in the pollen record. The GDD changed from that of 16 kya to that of the present at 500-year intervals; for example, at 16 kya GDD = 2,040; at 10 kya, GDD = 4,461; at 5 kya, GDD = 4,989; and at the present time, GDD = 5,395. The four simulation conditions were each run ten times, and the results were averaged. A smoothing technique was used to approximate the estimated 150-year duration of each pollen sample.

Figure 1A is the arboreal pollen diagram obtained from Anderson Pond. The simulated forest composition for the fourth condition—seed immigration and climate change—is shown in Figure 1B. The simulation using the third condition agreed with the pollen diagram almost as well. Conditions (1) and (2) resulted in simulations that were largely a mismatch with the pollen diagram. This indicated that only the simulations incorporating climate changes closely resembled the pollen record. Some discrepancies occurred. Ash entered later in the simulations than in the pollen record. This seems to have related to an ambiguity

(A)

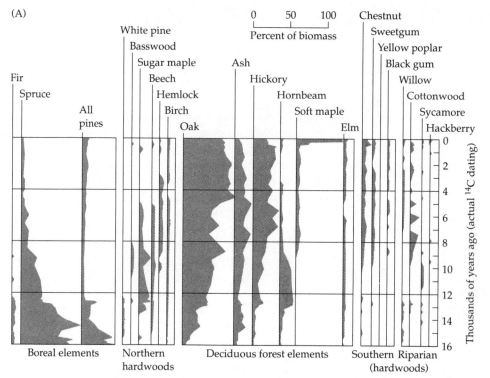

FIGURE 1 Actual forest succession, as determined from the pollen record, compared with a simulated succession. (A) Arboreal pollen diagram from Anderson Pond, Tennessee, for the past 16,000 years (based on carbon 14 dating). (B) FORET simulation of the past 16,000 years. This simulation was run under the fourth model condition—the climate changes, and seed sources are introduced into the model over time based on their first appearance in the actual pollen record. These conditions resulted in a simulated pattern that was substantially in agreement with actual succession. (From Solomon et al., 1981.)

as to just which ash species to use in the simulation. (In Chapter 4 it was mentioned that pollen records often cannot distinguish species within a genus.) Hickory did not appear in any of the simulation runs, but was a significant component of the pollen record, particularly during the last 8,000 years. Something clearly was wrong concerning the growth requirements for hickory used in the model. Also, the pollen record is not a perfect representation of the forest composition at the time, due to the size of Anderson Pond, sedimentation disturbance, and other factors.

A short commentary is in order here. In Chapter 4, in the discussion

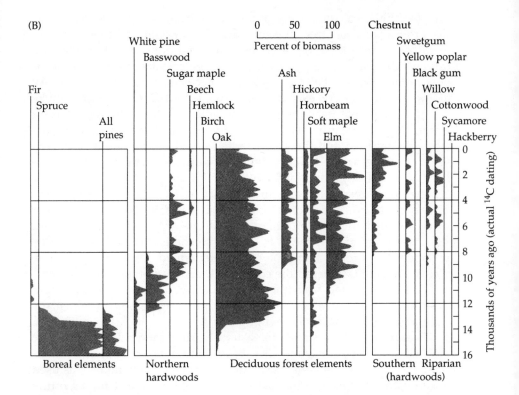

(B)

White pine
Basswood
Sugar maple
Beech
Hemlock
Birch
Oak

Fir
Spruce
All
pines

0   50   100
Percent of biomass

Ash
Hickory
Hornbeam
Soft maple
Elm

Chestnut
Sweetgum
Yellow poplar
Black gum
Willow
Cottonwood
Sycamore
Hackberry

Thousands of years ago (actual ^{14}C dating)

0
2
4
6
8
10
12
14
16

Boreal elements

Northern
hardwoods

Deciduous forest elements

Southern Riparian
(hardwoods)

of the rates of forest migration in North America, I quoted Davis (1981) to the effect that forest migration confirmed the "individualistic" concept of plant communities; that is, that each individual species acts independently of the other species. The model analysis of the Anderson Pond community appeared to contradict the Davis assertion and support the idea of the convergence of vegetation from different origins into a singular community structure—a distinctly Clementsian viewpoint. It is possible that both interpretations are correct. Davis was working with forest trees "on the move," where there were few dominant species and there was considerable lack of environmental stability. Solomon et al. (1981) were working with a forest where many species were available and their individual life histories combined with landscape stability produced communities of predictable species structure and composition.

The FORENA Model

Solomon (1986) developed another so-called gap model based on FORET, given the acronym FORENA (Forests of Eastern North America). FORENA calculated the annual establishment of new seedlings, the growth of each tree present, and the mortality of trees on one-twelfth hectare plots, all based on plot conditions. Light attenuation through the forest canopy, according to the calculated leaf area index, determined the degree of shading for species beneath the canopy. All of the routines were exactly like those of JABOWA. Hence, unlike FORET, individual trees competed for light, but they also competed for water. As stands grew larger and denser, the amount of nutrients available declined and the growth rate slowed. During each simulated year the model applied the shading, crowding, winter cold temperature limit, drought days, and GDD to the optimum growth each tree could achieve at its respective age and reduced it accordingly. Some of the physiological properties of trees mentioned in Chapter 3 were similar to those used here. Subroutines for GROW, SPROUT, GAUSS, and KILL were similar to those in FORET, with some refinements. Monthly precipitation entered a soil column of predefined depth and soil moisture capacity and then left the column either as runoff or as evapotranspiration, determined from the monthly temperature. Annual drought days were the number of days in each growing season in which the soil moisture was below the wilting point.

Twenty-one forest stands in eastern North America (Figure 2) were selected for simulation, and the appropriate data on each site's characteristics and species composition were included. The model was used to simulate forest composition as a function of time in a changing future climate. The following conditions were imposed: initially each plot was empty; a modern climate continued undisturbed for 400 years as the forest developed; 72 tree species were available to enter the plot at all times; at year 400, a 100-year-long linear climate change took place, ending with a climate equivalent to that expected under a doubling of present-day CO_2 concentration ($2 \times CO_2$); at year 500, a 200-year-long linear climate change took place, ending with a climate eqivalent to that expected under a quadrupling of present-day CO_2 concentration ($4 \times CO_2$); and at year 700, the climate stabilized for the next 300 years.

The climate scenarios in FORENA were from Mitchell (1983) and Mitchell and Lupton (1984), based on the GCM models by the Geophysics Fluid Dynamics Laboratory, Princeton. The climate changes included increased temperature changes with latitude. These ranged from 1°C in the south to 7.5°C in the north under $2 \times CO_2$; winter temperature increases ranged from 2°C in the south to 6°C in the north with $2 \times CO_2$ conditions and from 2.5°C to 7.5°C at $4 \times CO_2$; summer temperature increases ranged from 1°C to 3°C at $2 \times CO_2$ and from 2.5°C to 5°C at $4 \times CO_2$. The changes also included a summer precipitation decrease near the Great Plains, a summer precipitation increase along

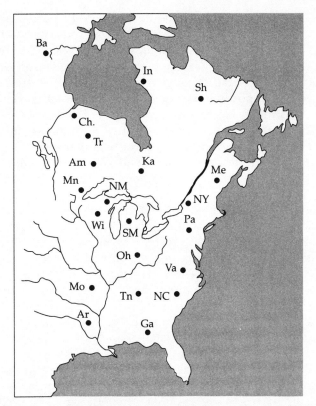

FIGURE 2   Locations used to simulate forest response to climate change with
FORENA: Baker Lake, Northwest Territories (Ba); Churchill, Manitoba (Ch);
Trout Lake, Ontario (Tr); Inoucdjouac, Quebec (In); Armstrong, Ontario (Am);
Kapuskasing, Ontario (Ka); Shefferville, Quebec (Sh); northeast Minnesota
(Mn); western Upper Michigan (NM); north-central Wisconsin (Wi); central
Lower Michigan (SM); northern Maine (Me); St. Lawrence Valley, New York
(NY); upper Susquehanna, Pennsylvania (Pa); west-central Ohio (Oh); eastern
Piedmont, Virginia (Va); southern Piedmont, North Carolina (NC); Cumber-
land Plateau, Tennessee (Tn); southwest Georgia (Ga); western Ozarks, Mis-
souri (Mo); and south-central Arkansas (Ar).

the East Coast, and no change in winter precipitation except for a
decrease in the southern United States. Monthly temperature and precipitation variation were not changed for the CO_2-induced climate
shifts. This is an important point, since warmer climates usually have
lower temperature variance and greater precipitation variance with decreasing amounts. However, Solomon et al. (1981) considered the data
basis for including this feature inadequate.

The FORENA model simulated forest composition at 5-year intervals for each of the 21 sites. Runs for each site were replicated 10 times and averaged. The model interpolated mean climate values for each year between 400 and 700 years, except that the growing season was shifted at 450, 550, and 750 years. The model was validated by testing it against modern forest composition data and by testing earlier versions against the pollen records in North America over the last 10,000 years and at 16 kya. The model outputs were consistent with long-term variations measured in forests, including long-term temporal sequences and spatial patterns, as reported by Solomon and Webb (1985). Other versions of the model were tested in North America, including Puerto Rico, and in Australia.

The model simulations of the modern distribution of forest ecosystems in North America were found to be generally correct. The simulations were physiognomically correct, showing forests where forests grow, nothing where trees are absent, and stunted woodland where scattered trees grow today. The 72 tree species introduced at 21 sites provided 1512 ways for the simulations to be wrong at the species level—that is, to show species growing where or when they should not, or not growing where or when they should, or growing in too great or too little abundance. A detailed analysis of the results yielded fewer than 150 such occurrences. Solomon et al. (1984) concluded that the model was reasonably valid, although not perfect. They wrote: ". . .the model must be viewed as being no more than an abstraction, and its picture of the future as only an hypothesis, not a prediction."

Simulated Forest Futures

The FORENA model simulations described above predict major ecosystem distribution changes with the projected greenhouse warming. Older, mature trees may die quickly, but their seedlings require decades or longer to become established, grow, and reach maturity. As the climate warms, many tree species will die back. Warmer-adapted tree species will slowly replace cooler-adapted species, but with considerable time lags. Let us now consider FORENA's projections for each of the major biomes in eastern North America.

Boreal Forests

FORENA simulated well the present tundra–woodland and boreal forests at the end of 400 years of growth. These were made up of black and white spruce with small amounts of birch and balsam fir entering in the southern parts. The model grew no trees on the modern tundra. The model did not allow for permafrost, fires, or possible terrain complexities.

Trees did not grow in the tundra at Inouecdjouac, Quebec, 300 km north of the present tree line along the eastern shore of Hudson Bay, even under the $4\times CO_2$ climate. On the west side of Hudson Bay, the $4\times CO_2$ climate was necessary for a closed forest canopy to develop at Baker Lake, Northwest Territories, 350 km north of today's tree line. The $2\times CO_2$ climate allowed closed spruce forest, with a small amount of paper birch, to replace what today is open parkland at Churchill, Manitoba and at Shefferville, Quebec. The climate was too cold and the growing season too short for most deciduous forest species to compete at these high latitudes. Under the $2\times CO_2$ climate at Trout Lake, Ontario (Figure 3A) there was an increase in total forest biomass as white spruce and paper birch grew more rapidly. Birch showed a cyclic behavior over time, which was related to its rapid growth, short life span, and the wide environmental tolerance of its seedlings. White spruce increased as conditions approached its temperature optimum, then declined from south to north as the $2\times CO_2$ climate became too warm. During the transition to the $4\times CO_2$ climate and beyond, the vegetation composition stabilized by year 700 (300 years from the present) at Kapuskasing, Ontario; by year 850 at the cooler, drier Armstrong; and not until after year 1000 at the coldest, driest site at Trout Lake. Sugar maple, for example, could grow at the southern sites early on, but not in the northernmost regions until much later. While the deciduous forest species were trying to get established, the boreal species were rapidly dying out, and were gone before the $4\times CO_2$ climate reached a steady state. In general, the simulated forest dynamics consisted of the rapid dieback of local species, followed by a gradual replacement by new species with a lag of 50 to 100 years. The actual migration rates of species were not used in the model; seedlings were made available at all times, but only when the climate became favorable did they become established.

Transition Forests

In the transition forests, deciduous hardwood species from the south mix with coniferous species from the north, each at the northern or southern limits of their respective ranges. FORENA simulated most of the community composition of today's transition forests quite well for sites in northern Maine, New York, Michigan, Wisconsin, and Minnesota. The 1000-year simulations for three sites in northern deciduous forests are shown in Figure 3B. During the transition to the $2\times CO_2$ climate there was a dieback of most species in the simulated coniferous-deciduous forests, then a recovery of many of the broadleaved species from years 500 to 600, followed by a second dieback to year 700 and a second recovery as community equilibrium was reached during the $4\times CO_2$ climate. In central lower Michigan, the warming climate promoted the growth of oaks, maples, ashes, white pine, and even walnut

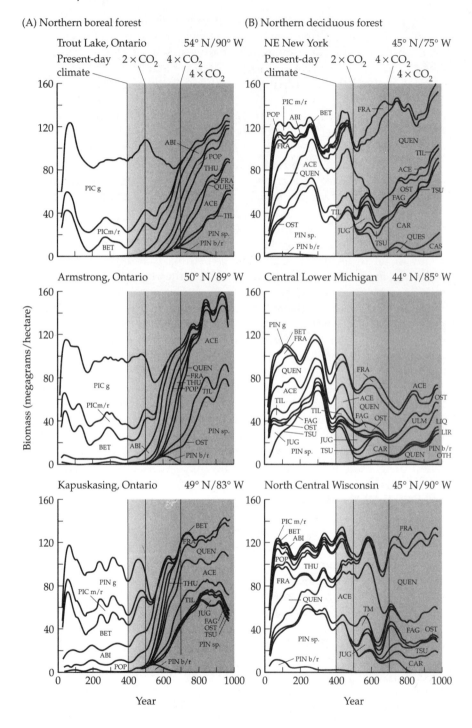

(A) Northern boreal forest

(B) Northern deciduous forest

◄ FIGURE 3 Simulations of 1000-year biomass dynamics, including CO_2-induced climate change, at three sites in the northern boreal forest (A) and three sites in the northern deciduous forest (B). Present-day climate conditions occur at year 400 of the simulation, with a doubling of atmospheric CO_2 concentration occurring at year 500. Tree taxon abbreviations: ABI, firs; ACE, maples; BET, birches; CAR, hickorys; CAS, chestnut; FAG, beech; FRA, ashes; JUG, butternut and walnut; LIQ, sweet gum; LIR, yellow poplar; OST, hornbeams; PIC, spruces; PIN, pines; POP, poplar and aspens; QUEN, northern oaks; THU, cedars and tamarack; TIL, basswood; TSU, hemlock; and ULM, elms. (From Solomon, 1986.)

until year 300, when a decline of all forest species set in from then through the 2×CO_2 and 4×CO_2 climates. Oaks dominated the 4×CO_2 climate transition period from year 500 to 700 and declined after that. Bitternut hickory and elms increased, along with oaks, from years 700 to 1000 in central lower Michigan. In northeastern New York, oaks and bitternut hickory dominated the forest from year 600 through establishment of the 4×CO_2 climate at year 700 and onward.

Figures 4 and 5 show the simulated sugar maple and white pine biomass dynamics at five western and five eastern sites in transition and boreal forests (Solomon et al. 1984). Sugar maple grew best during the transition to the 2×CO_2 climate in central lower Michigan, northeastern Pennsylvania, and northeastern New York and declined after that, but did well in north-central Wisconsin, northwestern Michigan, and northeastern Minnesota during the 4×CO_2 climate evolution. Note that sugar maple disappears entirely from central lower Michigan by year 500, at the end of the 2×CO_2 climate warming. Sugar maple would appear to do well in northern Maine and then in Ontario as the climate becomes warmer and warmer. White pine reached its peak biomass earlier than sugar maple in New York, Pennsylvania, central lower Michigan, and north-central Wisconsin, but declined strongly as the climate warmed beyond year 400. According to the simulation, white pine—already declining in central lower Michigan before the present—would not grow at the central to northern Ontario sites until the very much warmer 4×CO_2 climate had arrived.

Solomon and Bartlein (1992) used a forest gap model derived from FORET to simulate past and future forest ecosystems in northern Michigan. They used pollen records from three lakes in the western Upper Peninsula of Michigan to reconstruct the abundance of forest species there during the past 10,000 years. By the use of transfer functions they inferred paleoclimate conditions at these sites over that period. Using their FORET model, they simulated the vegetation sequence under the inferred climate. Once this had been done and a number of adjustments

(A)

(B)

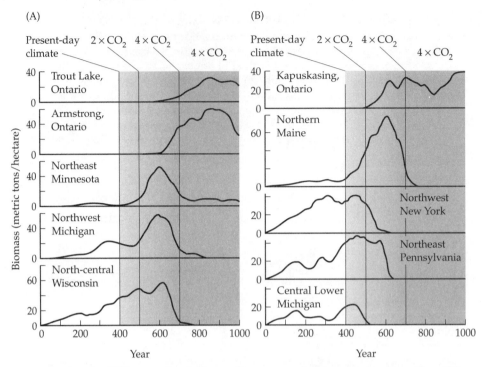

FIGURE 4 **1000-year simulations of sugar maple biomass incorporating CO_2-induced climate changes at five western (A) and five eastern (B) sites. (From Solomon et al., 1984.)**

made to the model, they simulated future forests starting with current conditions. They ran the model starting from a bare plot with 72 tree species available for a period of 400 years. This established the modern forest in the model. After year 400, the climate warmed linearly until the $2 \times CO_2$ climate was reached 100 years later, at which point summer temperatures had increased 2°C and winter temperatures 3°C, and there was 3.0 cm less January precipitation. The climate continued to warm until the $4 \times CO_2$ conditions were reached 200 years later (at year 700). Between years 400 and 500, as temperatures increased, several species died out, including boreal spruce, poplar, northern white cedar, and paper birch. The dieback reached its maximum by year 480. Sugar maple, northern oaks, and ash increased and peaked at year 600, declining after that. Biomass increased after year 700 as basswood, beech, hemlock, and bur oak invaded, but sugar maple and white pine almost disappeared.

The warming projected in the model for the $2 \times CO_2$ climate within 100 years was equivalent to an increase of 500 GDD. Compare this with the proxy warming during mid-postglacial times, which took 2000 years

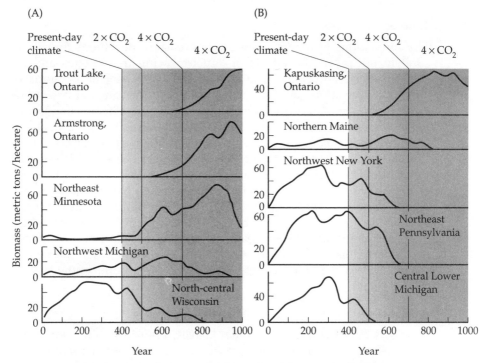

FIGURE 5 **1000-year simulations of white pine biomass incorporating CO₂-induced climate changes at five western (A) and five eastern (B) sites. (From Solomon et al., 1984.)**

to produce a 125 GDD increase. At the projected rate of warming, this much warming would occur in only 25 years. Solomon and Bartlein (1992) also modeled the effects of CO_2 enrichment on forest tree species. They wrote:

The most striking result of the simulated CO_2 effect was to compensate for losses of biomass during the initial dieback. In addition, the entire progression of forest communities occurred about 60 years earlier than in simulations lacking CO_2 effects. Unlike such simulations in adjacent boreal forests and temperate deciduous forests, stand biomass at the northern Michigan study area increased markedly, a change induced by more rapid tree growth in gaps, and by reduced mortality rates in this peculiarly sensitive forest flora.

The authors went on to note that some of the species that were successful in the simulation may be too far away to occupy the niches that become available in northwestern Michigan as the climate warms—their migration rates are just too slow. For example, beech now grows 60 km (300–430 years away) to the east, white oak is 85 km (200 years away) to the south, and chinquapin oak is 225 km (550 years) away. Their presence in northwestern Michigan would require human intervention.

Temperate Forests

The simulations of modern western and eastern deciduous forests accurately portrayed the effects of larger-growing tree species and increased species diversity. The simulations were realistic in terms of species composition and dynamics throughout the range of deciduous forests. The simulated response of these forests to climate change was remarkably uniform. Biomass declined at all sites after the $2 \times CO_2$ climate arrived, except in south-central Missouri and Arkansas. There biomass decline began with the warming at year 400 as forest reverted to prairie. These sites are xeric (dry) at the present time and can be characterized as forests with prairie openings. In North Carolina, Virginia, Tennessee, and west-central Ohio, biomass declined as the result of moisture stress (days of soil moisture below the wilting point), which increased eightfold between years 400 and 700. Moisture stress reduced the stand biomass by reducing the potential growth rates of the rapidly growing species, giving a competitive advantage to some of the slower-growing species.

A Model with Nitrogen and Carbon

Nutrients cycle within ecosystems. Leaves, branches, and even whole trees fall to the ground, decompose, and return organic matter and nutrients to the soil. The carbon and nitrogen cycles are strongly and reciprocally linked. The amount of carbon and nutrients in fallen plant material differs greatly among species and even within species. The lignin content of leaves affects the carbon chemistry of soils through leaf litter. A high lignin content (more carbon) means less breakdown of the leaf litter, less available nitrogen, and a slower nitrogen cycle. Water availability is also a major factor controlling tree growth because it is the flow of water through a tree that carries nitrogen, carbon, and other nutrients to its parts.

Pastor and Post (1986, 1988) examined the sensitivity of forest responses to the carbon, nitrogen, and water cycles using a gap dynamics stand model (basically FORENA) to simulate 11 sites in Canada and the United States. The model was run for 200 years under current climate, which then warmed linearly over the next 100 years to reach the $2 \times CO_2$ climate and then was run for 200 more years with the new climate at steady state. At Shefferville in northern Quebec, spruce productivity and biomass increased over time and the boreal forest was not replaced with deciduous hardwoods. At sites in the southern boreal and northern hardwood forests from the western Great Lakes to Maine, water balance had a major effect on simulated species composition and carbon balance. Where there was no decrease in soil water availability with the $2 \times CO_2$

climate, the mixed spruce–fir–northern hardwood forest was replaced by a more productive hardwood (sugar maple and paper birch) forest. The hardwoods have a higher intrinsic growth rate than either spruce or fir, and the higher nitrogen and lower lignin content of hardwood litter makes for increased nitrogen availability. This enhancement magnifies the effect of warming on productivity when water is available. On sandy, well-drained soils, however, soil water dropped below the wilting point for a greater proportion of the growing season under the $2\times CO_2$ climate. Spruce–fir–northern hardwood forests from Minnesota to Maine were replaced by a stunted pine–oak forest of much lower carbon content. This also happened to a lesser extent on the silty clay loam soils of northern Wisconsin, southern Michigan, and northern New York. As Pastor and Post wrote, "Thus, response of boreal and north temperate forests to CO_2-induced climatic change may depend on the balance between changes in the hydrologic cycle that constrain the forest response, and the positive feedbacks between the carbon and nitrogen cycles that amplify this response."

The Alaskan Boreal Forest

High-latitude regions are expected to experience the greatest temperature increases with global warming and the arctic ecosystems may be affected the most. Changes in arctic ecosystems may represent early warnings of slower and more widespread global effects. Van Cleve et al. (1991) discuss element cycling in the boreal forest; their paper is an excellent introduction to the following models of the boreal forest, as are the works by Shaver et al. (1992) and Chapin (1992).

Approximately 31% of Alaska is forested, and the remainder is a mosaic of grassland, shrub, bog, and tundra. In the interior uplands of Alaska, precipitation changes may have a greater influence on forest succession and growth than do any changes in air temperature. Here the interactions among soil temperature, soil moisture, forest floor organic matter, and fire history largely control forest productivity and nutrient cycling. Black spruce grows here, mainly on cold, wet, nutrient-poor, north-facing and bottomland sites with permafrost. White spruce, paper birch, and trembling aspen grow in more productive stands on warm, well-drained, south-facing and bottomland slopes without permafrost. If the climate of Alaska were to become warmer and drier in the future, it was suggested by Viereck and Van Cleve (1984) that steppe-like vegetation and trembling aspen stands would expand on dry sites and paper birch would move into wet areas now occupied by black spruce. Emanuel et al. (1985b) predicted that dry boreal forests would evolve into steppe-like vegetation.

Bonan et al. (1989) used a forest gap model to examine the sensitivity of permafrost and permafrost-free forests in interior Alaska to air temperature and precipitation changes. They added several features to the forest gap models used earlier, such as depth of seasonal soil thawing, tree mortality as a function of stress, forest fires, and insect outbreaks, in addition to longevity. They also modeled moss–organic accumulation on the forest floor as a function of moss productivity and decomposition, depending on the site conditions. Fire severity was a function of fuel buildup and forest floor moisture. The two types of sites mentioned in the previous paragraph, black spruce and white spruce–hardwood sites, were modeled. A variety of steady-state climate scenarios were used, encompassing mean monthly air temperature increases of 1°C, 3°C, and 5°C; and mean monthly precipitation increases of 120%, 140%, and 260%. A short-term model run considered the effects of these climate changes on forest succession for 200 to 250 years following a fire. A long-term run modeled the effects of climate changes on equilibrium forest patterns for 500 years with recurring, random wildfires, giving an estimated equilibrium forest stand composition and size.

Black spruce permafrost forests   In this model, a catastrophic fire occurred initially only, consuming the forest floor to a depth of 10 cm. Seeds from all species were available to germinate immediately following the fire. (Recall from Chapter 3 that fire is essential in order for the seeds of some conifers to germinate.) Only black spruce was able to grow on this site under all the short-term climate conditions. The black spruce stand grew optimally with the temperature 3°C above average present-day levels, but grew less under the 1°C and 5°C increases. The increased potential evapotranspiration resulting from warmer air temperatures dried the soil organic layer, reduced the thermal conductivity of the forest floor, and decreased the depth of seasonal soil thawing, all of which limited forest growth. However, when precipitation was allowed to increase, this offset the increased potential evapotranspiration, the depth of seasonal soil thawing increased considerably, and simulated stand biomass increased. Greatest stand biomass accumulation occurred with a precipitation increase of 160% under the 1°C and 3°C warmings.

The long-term (500 years) simulation under present climate conditions resulted in a forest of dominant black spruce, and to a lesser extent paper birch, with white spruce and trembling aspen as minor components. Under the 1°C warming, black spruce continued to dominate. When precipitation did not increase, the paper birch biomass doubled from the baseline amount. Increases in precipitation increased black spruce biomass and decreased paper birch biomass. Under the 3°C warming scenario, paper birch became the dominant tree, with black and white spruce of secondary significance. When precipitation did not increase, paper birch and trembling aspen biomass quadrupled from the baseline climate and white spruce, although of secondary importance, increased its biomass tenfold. Precipitation increases had little effect on stand structure other than to increase the black spruce biomass. Under the 5°C warming, forest composition shifted and the hardwoods dominated. Trembling aspen biomass increased tenfold and paper birch biomass sixfold from the baseline condition, while black spruce decreased. Increased evapotranspiration produced a drier forest floor moss–organic layer, a greater depth of fire burn, an increased depth of seasonal soil thawing, and an increased abundance of hardwood species and white spruce.

White spruce-hardwood forests The short-term sensitivity simulation to climate involved an initial fire and then 250 years without other fires. Seeds of white spruce, paper birch, and trembling aspen were available. Under present-day conditions, the simulated forest was initially dominated by paper birch, which was succeeded by white spruce as the birch canopy died back. Observed successional changes match this simulation. Under the 3°C and 5°C climatic warmings, none of the tree species simulated were able to grow. The strong evapotranspiration demand associated with the warmings dried the mineral soil below the drought tolerances of these tree species (30% for paper birch and white spruce, 40% for trembling aspen). Tree species did grow successfully with the 1°C warming. If precipitation did not change, trembling aspen dominated the forest. When precipitation offset the increased potential evapotranspiration demand, the forest became a mesic paper birch–white spruce stand. Under the long-term (500-year) forest simulation, the structure and composition of the forest depended on the fire regime. A warmer, drier climate increased the frequency of fire, which in turn set the forest back to early stages of succession. A shorter fire cycle increased the abundance of paper birch and trembling aspen over that of white spruce.

The forest gap model used here clearly showed that precipitation and potential evapotranspiration had a greater influence on the Alaskan boreal forest biomass than did the climate warming directly. On poorly drained, north-facing slopes, climate warming allowed only black spruce to dominate for all climatic scenarios in the shorter term (250 years) without recurring fires, but in the longer term (500 years) with frequent fires, black spruce forests converted to mixed hardwood–white spruce forests. On well-drained, south-facing slopes, white spruce–hardwood stands were converted to dry aspen forests or even to steppe-like, treeless vegetation.

Moss–forest interactions Many of the coniferous boreal forests of northern North America, Scandinavia, and Eurasia have a thick layer of moss and decaying organic matter on the forest floor. This thick moss–organic layer precludes the regeneration of many species of pine, birch, spruce, and aspen, although a few tree species do regenerate successfully in such a layer. Since mosses thrive in moist, shaded habitats, an open forest canopy leads to a decline of moss cover, although at the other extreme a dense forest canopy will shade out the mosses.

According to Bonan and Shugart (1989), a moss layer controls energy flow, nutrient cycling, water relations, stand productivity, and stand dynamics. The low density and low thermal conductivity of a moss layer insulates the underlying mineral soil, lowers soil temperatures, helps maintain permafrost, impedes soil drainage, promotes slow tree growth, and slows organic matter decomposition. Mosses have a high capacity to absorb water and to hold nutrients until the mosses die. When the mosses decompose, the nutrients are released and become available to vascular plants. However, an undecomposed organic layer may retain nutrients indefinitely. With global warming, there may be increased moss–organic layer decomposition and a significant release of nutrients to the boreal forest ecosystem. Bonan (1991) improved on the earlier biophysical model and found that slope and aspect (north- vs. south-facing) had little influence on soil temperature, but that the effects of forest canopy and moss layer cover were very important.

Prentice et al. (1989) developed a new algorithm based on a more detailed physiological understanding of moss dynamics for the boreal forest model originated by Bonan and Shugart (1989). Their early model results show a wide range of moss–forest interactions, depending on the site conditions. Thick moss layers develop on cold, wet sites that exclude trees, while on warmer sites a thick moss layer still controls the vegetation but allows spruce to grow. On warm, moist sites, birch and aspen prevent the development of a moss layer. However, as these trees die and are replaced by longer-lived, shade-tolerant spruce, a thin moss layer forms. If conditions are warm and dry, the forest is dominated by birch and aspen without mosses. It will be important to apply this model to Klinger's ideas concerning paludification as described in Chapter 2.

FORSKA: A Boreal Forest Gap Model

Prentice et al. (1991) have developed a forest gap model that is an advance over earlier models in details of physiological functions and tree structure. Earlier models used a parabolic function for the response of growth to annual accumulated temperature sums. However, evidence suggests that annual growth of trees should not decline with increasing

GDD, as is given by the parabolic function. In the FORSKA model developed by Prentice et al., growth continues to increase as GDD increase. FORSKA uses a parabolic function for growth response to drought, ranging from unity when soil moisture is adequate to zero at the species' limits. This model uses vertically distributed tree crowns, which better represent boreal conifers than do disc models with leaves concentrated at the top, shading other shorter trees beneath them. The earlier models worked well for the deciduous forests and reasonably well for some conifer stands.

In FORSKA, growth and regeneration are suppressed if temperature and drought functions are zero, or if the temperature of the coldest month is too low, or if the GDD sums are not adequate for the species. For example, for Scandinavian trees, regeneration is zero if the warmest-month temperature is below 16.0°C for the lime tree, *Tilia cordata*, 16.5°C for beech, *Fagus sylvatica*, and 17.0°C for hornbeam, *Carpinus betulus*. Regeneration is also zero if the coldest-month temperature is above −1.5°C for Norway spruce, *Picea abies*, or −2.5°C for gray alder, *Alnus incana*. No cold winter requirement is imposed on Scots pine, *Pinus sylvestris*.

Prentice et al. (1991) used the FORSKA model to simulate the response of forest vegetation to global warming on a transect in Sweden across the ecotone that separates the boreal coniferous and temperate deciduous forests. The poleward limits of all tree species here are related to growing season warmth, with each species having its own GDD requirement. The equatorward limits are controlled by drought, by winters too mild to meet chilling requirements, and by competition with temperate species. All 20 tree species native to the region were included in the model. The landscape was treated as an array of patches, which were subjected to a disturbance treated as a stochastic process increasing with time and repeated with an average 100-year return time. Each set of patches was run under the present climate for 400 years. The forest composition and biomass after 400 years of simulation closely resembled that of the modern forest. Monthly temperature and/or precipitation were then increased linearly over 100 years to fit GCM projections for the $2\times CO_2$ climate. The forest model was allowed to equilibrate in the $2\times CO_2$ climate for a further 300 years. Temperature increases were in the range of 2°C to 4°C in summer and 5°C to 6°C in winter under greenhouse warming.

With increases in both temperature and precipitation (175 to 250 mm per year), the present-day northern boreal forest changed to a transition forest with small amounts of English oak (*Quercus robur*) and other temperate species. The central, transition forest zone, showed a greater change, from dominance by spruce to a forest of beech, oaks, and pine, lacking spruce. The southern temperate deciduous forest showed relatively little change, although the abundance of beech increased. Increasing precipitation alone had no effect at the northern zone, but did increase biomass further south, particularly of spruce and beech. Drought seemed to limit the growth of these species in the transition zone. Increasing temperature with no change in precipitation produced drought in the transition and temperate forest zones, resulting in a biomass gain less than that made under increasing precipitation. Pines were favored under these circumstances. The transitions described under temperature and precipitation increases took 150 to 200 years to occur. Many of the forest changes occurred after the climate change was complete, a lag caused by ecological processes.

A Longer-Term Model

The forest gap models described above give us considerable information about the responses of individual species and communities of species on a year-by-year basis over decades to hundreds of years in response to climate change projected by GCMs. There is a need for broad landscape vegetation models of coarser time resolution which can run with longer time horizons (decades, centuries, and even millenia) and be interactive with longer-term GCMs. Prentice et al. (1989) are developing a long-term model they call a dynamic global vegetation model (GVM). This is "a model of the dynamics of the soil–plant–atmosphere system on time scales of 10 to 1000 years that includes all the critical ecosystem processes—physical, chemical, and biological—operating on this time scale."

The GVM incorporates a DYNAMIC stand simulator that works at the taxonomic resolution of plant functional types as defined by Huston and Smith (1987) to yield time-dependent patterns of plant growth, mineral cycling, water use, etc., and a SPATIAL static grid model made up of a coarse regional grid containing restraints to growth such as monthly temperatures and precipitation, and including actual and potential evapotranspiration. Also included is the character and variation of soil texture, water-holding capacity, and fertility; topographic diversity; solar insolation; and latitude effects. The DYNAMIC stand simulator is nested within each grid cell of the SPATIAL static model. The coarse grid system contains a definition of climate-related life zones for vegetation and crops. The dynamics of natural and managed ecosystems can be modeled stochastically at the landscape level by simulating a large number of vegetation patches (1000 m^2). This entire scheme is still under development and may require quite some time before it produces definitive results.

Other Forest Gap Modeling

The U.S. Environmental Protection Agency (see Smith and Tirpak, 1989) sponsored several models of forest ecosystems within the boundaries of the contiguous United States. The EPA specified that the climate scenarios used must be the outputs from the following GCMs: Goddard Institute of Space Studies (GISS) (Hansen et al., 1988); Geophysical Fluid Dynamics Laboratory (GFDL) (Manabe and Wetherald, 1987); and Oregon State University (OSU) (Schlesinger and Zhao, 1988). The scenarios used were actually hybrids between GCM average monthly estimates of temperature, precipitation, and other weather variables and daily historic weather records. Monthly values from model simulations of $2 \times CO_2$ climate conditions were divided by those from model simulations of current conditions in each grid box. These ratios were then multiplied by the actual mean monthly temperature and precipitation to generate "treatment" climates.

All three GCMs projected the steady-state climate for the equivalent of double the current atmospheric CO_2 concentration in about 2060. They each estimated that average temperatures over the United States would rise, but they differed as to the magnitude: OSU projected 3°C, GISS 4.3°C, and GFDL 5.1°C. The seasonal patterns predicted by the three models were different, as seen in Figure 6. GISS had the largest warming in autumn and winter, GFDL had the largest in spring, and OSU had little seasonality. All three models estimated that annual precipitation over the northern United States would increase. The GISS and OSU models showed increases of 73 and 62 mm respectively in annual precipitation, while GFDL showed 33 mm. The first two models showed precipitation increases throughout the year, but GFDL showed a decline in summer rainfall.

In addition to its steady-state climate modeling, the Goddard Institute for Space Studies modeled how global climate may change as concentrations of greenhouse gases gradually increase during the next century. GISS transient model A assumed that trace gases continued to increase at historic rates, which have been exponential, and that the net greenhouse forcing of climate increased exponentially. By the end of this scenario transient model A showed global warming equivalent to that under the equilibrium $2 \times CO_2$ climate. GISS transient model B assumed a decreasing greenhouse gas concentration growth rate such that climate forcing increased linearly and stopped altogether in year 2029. GISS model B included possible volcanos. Decadal average temperature changes for both GISS transient scenarios are shown in Figure 7.

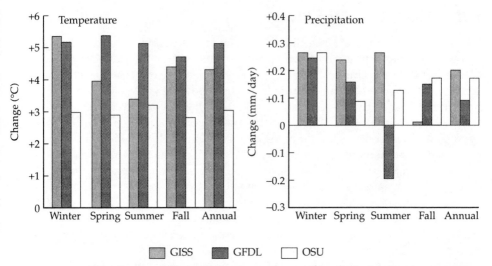

FIGURE 6   **Average changes in temperature and precipitation from present-day conditions (0) over the contiguous United States as projected by three GCMs for the 2×CO₂ climate. (From Smith and Tirpak, 1989.)**

Great Lakes Forests

Botkin et al. (1989) used a modified JABOWA model that they called TIMBER! along with JABOWA II to simulate the effects of climate change on Great Lakes forests. They used the climate outputs of the GCMs described above to drive these models. The JABOWA II model incorporated the following improvements over JABOWA: a more complete way of handling the relationship between water and tree growth in order to distinguish between floodplain communities and bog/wetland communities; a relationship between soil nitrogen and tree growth; and a more realistic treatment of growth and reproductive rates among species. The model incorporated 40 tree species characteristic of the Great Lakes region. Two representative sites were used. One site was in the northern part of the Great Lakes region near Virginia, Minnesota, and included the heavily forested lands of the Superior National Forest and the Boundary Waters Canoe Area. This area is a transition zone between northern hardwoods and boreal forest. The dominant species are spruce and fir on wetter soils, and aspen, white birch, and pines on drier sites. The second site was in the southern part of the Great Lakes region near Mount Pleasant, Michigan, an area characterized as transitional between northern hardwoods, beech, and sugar maple to the north, and oak-dominated forests to the south.

Model experiments were conducted on the growth of these Great

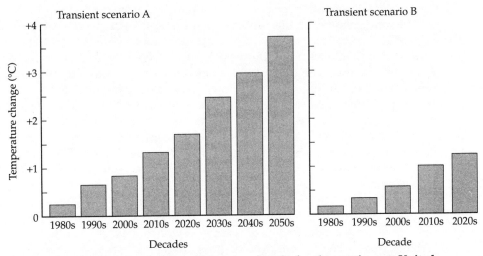

FIGURE 7 **Average temperature change per decade for the contiguous United States as projected the GISS transient A and transient B models. (From Smith and Tirpak, 1989.)**

Lakes forests from an original clearing under the following climate scenarios: (1) the 1951–1980 climate; (2) GISS steady state; (3) GFDL steady state; (4) OSU steady state; (5) GISS transient A; and (6) GISS transient B. The forest model was run with each climate scenario for 100 years (except for the transient A model, which was run for 90 years). Each experiment was replicated 60 times, the results were averaged, and the 95% confidence intervals were calculated. The following soil conditions were used: (1) deep, comparatively dry soils; (2) deep, comparatively wet soils; (3) shallow wetland soils; and (4) shallow, dry, upland soils. In addition to the regrowth on clear-cut plots, in Minnesota the effect on old-growth (400-year-old) forest stands was also considered.

Under all the $2 \times CO_2$ scenarios, the change from the present-day climate was dramatic. The Mt. Pleasant, Michigan climate went from winters that have months with mean temperatures below freezing to a series of years with no months below freezing. The current mean maximum January temperature became the mean minimum January temperature. Similar increases occurred for the mean July temperature; currently it varies from 20°C to 25°C, but it changed to 28°C to 32°C or more. These temperature changes resulted in much more evapotranspiration from forest trees and in drier soils. Calculated precipitation did not change a great deal, but some years were drier. The researchers found similar warming increments for the Minnesota site.

The northern Minnesota forests   The transient climate scenarios applied to the forest model of the Minnesota site resulted in rapid changes, largely caused by the sensitivity of the natural forests of northern Minnesota to soil moisture. By the year 2010, a 400-year-old stand dominated by balsam fir on deep, fertile, moist soil declined to one-third of its current balsam fir basal area. However, sugar maple replaced fir as the dominant species and the total biomass nearly tripled. In a wetland, a 400-year-old white cedar forest declined to nearly treeless bog. On a deep, drier but fertile upland soil the dominant white birch declined about 90% in 40 years and was also replaced by sugar maple. The total standing biomass declined about 50% after 90 years under the $2 \times CO_2$ climate compared with what it would have been with the present climate over that time period. On deep, relatively moist soils, wetlands that now develop into larch-dominated bogs regrew to red maple-dominated wetlands characteristic of warmer regions today. A shallow, dry, upland soil, once cleared, would grow trembling aspen (a nitrogen-fixer) today, but this species was replaced by sugar maple with the long-term warming.

In the forests of the Boundary Waters Canoe Area of northern Minnesota and of Isle Royale National Park, such a transition from boreal to northern hardwoods would result in massive diebacks of mature trees during the coming century. The understory of these forests would become thickets of shrubs, seedlings, and saplings. The canopy trees, stressed by too much heat and water loss, would become susceptible to insect and disease outbreaks. The incidence of fires would increase as the dying trees provide abundant fuel. Botkin et al. (1989) write:

As a consequence of these major changes in vegetation, large alterations would occur in the forest ecosystems; chemical cycling, storage of organic matter, and rates of decomposition differ between conifer-dominated boreal forests and the northern hardwood forests. The habitat for wildlife would be altered, and one would expect the dominant species of wildlife to change with the vegetation. For example, areas suitable today to moose would become favorable to white-tailed deer.

The Mount Pleasant forests   The forest simulation of the Mount Pleasant, Michigan site under the actual climate for 1951–1980 showed red oak, sugar maple, white oak, and white pine as dominants on drier sites, and sugar maple, red maple, white ash, and basswood as dominants on wetter sites throughout the first 100 years of growth on a clear-cut area. Under the GISS transient A climate over 90 years, sugar maple did not occur on the drier sites, which were dominated first by pin cherry, then later by red and white oak and red maple. Sugar maple developed on the wetter sites during the first decades but then declined to near zero after 90 years. Under the GISS transient B climate over 60 years, sugar maple on wetter sites followed the same pattern as it did under the transient A climate. On the drier sites, forest productivity during the first 50 years of regrowth following clear-cutting under the transient A climate matched current productivity, while under the transient B climate productivity exceeded the current level for the first 50 years. However, as the climate warming effects increased, there was a decline in productivity by the year 2040 for both transients A and B, with biomass reaching a very low value by year 2070. The steady-state models produced similar projections, but with some differences among them. For dry soil conditions, the models all showed a shift from a northern hardwood–oak transition forest to an oak forest with red maple present. The GISS $2 \times CO_2$ steady-state climate led to an open savanna forest type with low biomass. The GFDL $2 \times CO_2$ steady-state model showed the forest being converted to sparse savanna or grassland with small trees.

On deep, well-watered sites with sandy soils having a well-aerated layer above a saturated zone, the forest under current conditions at Mount Pleasant is transitional, with sugar maple and wetland species such as white ash and hemlock. Red maple grows under disturbed situations. The GISS $2 \times CO_2$ steady-state climate scenario resulted in a forest dominated by red maple (thought unlikely by some modelers) and with oak present. Under the 1951–1980 actual climate simulation, sugar maple increased in biomass throughout the first 100 years, while no sugar maple grew at the end of each steady-state $2 \times CO_2$ climate scenario. Under the GISS transient A climate, sugar maple declined to disappearance before the end of 90 years. In the GISS transient B climate, sugar maple remained with little change in biomass at the end of 60 years.

The changes expected in central Michigan forests under the greenhouse warming expected during the coming century are enormous, particularly under drier conditions. Stands of sugar maple, yellow birch, and white pine will convert to forests of red and white oak and red maple. Botkin and colleagues did not introduce any southern forest species into the model, so there was no oportunity for warm-adapted species to grow. The implications of all of this for the commercial forest production of Michigan in the future is very great indeed.

Kirtland's warbler, an endangered bird, nests in the fire-controlled jack pine forests of north central Michigan. Botkin et al. (1991) performed an analysis of this special forest under climate change using the GISS transient A scenario and a JABOWA forest growth model. The models showed a huge potential decline of the jack pine forest under climate warming and a loss of Kirtland's warbler nesting habitat during the coming decades. The short-term response of the jack pine forest was dramatic because these trees currently grow at the southern limit of their range in north central Michigan. (See Chapter 6 for further details.)

Forests of the Southeastern United States

Urban and Shugart (1989) used a gap model derived from the FORET model of Shugart and West (1977), called ZELIG, to simulate the response of the upland forests of the southeastern United States to climate change. Present-day forests in the southern Appalachian Mountains are oak–pine, oak–chestnut, and shortleaf pine. Through the Piedmont and onto the coastal plain, the main trees are sweet gum, black gum, southern oaks, and loblolly pine. The southern pine forests, dominated by shortleaf and loblolly pines, are by far the most important commercial forests in the eastern United States.

Urban and Shugart modeled forests at four geographic locations: Knoxville, Tennessee; Florence, South Carolina; Macon, Georgia; and Vicksburg, Mississippi. The climate scenarios they used were the same as those used by Botkin et al. (1989), except that GISS transient B was not used. To create a baseline climate scenario, the 30-year weather record for each location was concatenated to yield a scenario of sufficient length to give a meaningful forest response. Simulations of the baseline climate and each of the three steady-state $2 \times CO_2$ climates were then run, each 200 years. The first few decades of each run represented successional forest stages, while the later years indicated trends of more mature forests. The results reported represented the averages of 50 replicate model plots.

In contrast to the precipitation increases projected for the Great Lakes region, the model simulations all showed a decrease in moisture for the southeastern United States with the warming climate. All three $2 \times CO_2$ scenarios projected a substantial increase in growing degree-days for all the sites. At Knoxville the GDD increased from 3616 under the baseline climate to 4917–5327 for the $2 \times CO_2$ scenarios; at Macon, 4682 increased to 6075–6488. The decimal fraction of drought days out of the number of growing season days went from 0 at Knoxville under the baseline climate to 0.05–0.48 under the $2 \times CO_2$ scenarios; at Macon the increase was from 0.09 to 0.33–0.54.

None of the tree species simulated in the model could tolerate a drought-day index greater than 0.6, and only a few of the species could tolerate greater than 0.5. All species showed significant growth reductions if the drought-day index was greater than 0.3. A drought-day index of 0.5—that is, half the growing season under drought conditions—corresponds approximately to the western margin of the eastern deciduous forest today. Very few tree species in the model could survive under temperature regimes of more than 6000 GDD. The 5500 GDD isopleth today approximately parallels the Gulf Coast. Tropical species were not allowed to enter the model, and the maximum GDD limits for southern tree species are not well estimated, since the Gulf of Mexico is a geographical constraint to their southern range. So, in the model only marginal forests grew where GDD sums exceeded 6000.

Under the $2 \times CO_2$ scenarios in eastern Tennessee, the simulated succession for upland oak–pine forests was at first strongly dominated by shortleaf pine, which was then replaced by shade-tolerant hardwoods including black gum, elm, and Shumard oak. The GFDL scenario (which predicted the most severe warming and drying) resulted in no trees surviving and a grassland or sparse savanna developing. In the northern coastal plain of South Carolina, loblolly pine strongly dominated the early succession on upland sites under the baseline climate. Under the GISS scenario, loblolly pine was present but never common, and shortleaf pine was common but not dominant. Southern hardwoods occurred throughout the entire simulation; these included black gum, hackberry, laurel oak, and elm. At the end of the warming, these forests supported less than half the biomass of the baseline forest.

The upland sites in Georgia were dominated by loblolly pine; then saplings of southern coastal-plain species became established, but none persisted to develop a forest. Temperature seemed to be the primary limiting factor in their devlopment under the GISS climate. Under the OSU climate regime only xeric (drought-adapted) species, such as post oak and blackjack oak, started to grow, but they did not survive. This is because the model used the earlier age-independent mortality routine and as a result, species adapted to growing very slowly (such as post oak) were "killed" at the same rate as species not capable of slow growth (scarlet oak, for example). Drought seemed to be a limiting factor in this simulation.

Urban and Shugart then simulated a bottomland site in Georgia with a moderate regime of water-saturated soil conditions. A forest of black gum, sweet gum, and elms developed under the baseline climate; that is exactly what exists there today. Then the effects of temperature and summer droughts did their damage as the climate warmed, and those species dropped out.

In Mississippi, the successional stages were represented primarily by loblolly pine, except at drier sites which became post oak–savannas. Under all climate scenarios, the sites became too hot and dry to support trees.

The forests at the four sites were also modeled using the GISS transient A scenario for climate change. The 30-year weather data base for each location was again concatenated into the GISS transient A scenario to yield a 90-year transient scenario. Two cases were simulated for each location: a 90-year successional sequence from bare ground, and a 90-year projection of a 100-year old stand, which was the output of the baseline climate at year 100. In both cases the forests tracked the baseline-climate control forest until about 60 years. Then each modeled forest underwent a decline in standing biomass. In the successional forest, the decline was caused by a decrease in the abundance of the dominant shortleaf pine. The temperature, and therefore the GDDs, increased considerably at about year 60. In the mature forest there was a pronounced decline in standing biomass between years 60 and 70 (years 160 and 170 of the total simulation). Again, the decline was caused by the sudden warming occuring during this decade. During this period there was a breakup of the canopy and regeneration of some species. The decline resulted not just from the loss of the dominant species—shortleaf pine—but from the loss of several co-dominant species as well. Loblolly pine, which was favored under the GISS $2 \times CO_2$ scenario, did not appear in the transient forest until after year 70, and was still a relatively minor species at year 90.

Under the GISS transient A scenario in South Carolina, the climate had several strong warming spikes at 25, 55, and 85 years, each of less than a decade duration. In the successional forest there was a decline in standing biomass after 55 years followed by a steeper decline after about 80 years. In the mature forest two declines were evident after 55 and 80 years of warming, the first decline being small but the second quite abrupt. The transient climate for Georgia showed some warming after 25 years, but had a dramatic increase in mean annual temperature of 4°C to 5°C after 60 years. Total biomass declined abruptly after 60 years in both the successional and the mature forests. Many species declined, including loblolly pine. At the Mississippi site, the forests (both successional and mature) began to decline with warming after 20 years or so, and a dramatic decline occurred after 60 years, at which time the GDD exceeded 6000. This was the upper limit used in the simulation, after which the trees succumbed to heat stress. A similar point was reached at the South Carolina site after 80 years.

Urban and Shugart caution against believing all the results shown since what may happen in reality contains many unknowns. They predict the southern forests will degrade to marginal forest or nonforest vegetation, but the basis for this projection is uncertain. In real forests one would expect weather anomalies to modulate forest dynamics, with stressful periods resulting in episodes of heavy mortality. We might expect to see significant forest decline 40 to 70 years from now, but it is impossible to predict the precise timing of such decline.

Forest Sensitivity to Temperature Change

How large a change of temperature is necessary in order for there to be an observable change in forest biomass or composition? How long must a temperature change last in order for there to be a detectable change in forest biomass or composition? Davis and Botkin (1985) explored these questions by applying the JABOWA forest model to the cool-temperate forests of New England. They simulated the responses of these forests to temperature changes similar to those experienced in the Northern Hemisphere during the last few centuries. The questions they were asking were directed at the interpretation of past climates as inferred from the pollen record. Davis and Botkin wanted to know how small a temperature change might be derived from a change in pollen deposition. The methodology they used was similar to that described above for JABOWA but with some modifications to incorporate a more precise calculation of hydrological factors, to include the dependence of tree growth on soil fertility, and to allow several patterns of climate change.

The Little Ice Age Simulation

A 2°C change in mean annual temperature approximates the largest change observed at temperate latitudes in the North Atlantic region during the past 500 years: the Little Ice Age. Davis and Botkin approximated the Little Ice Age as a 200-year interval that cooled by 600 GDD. In their model both the amplitude and the length of the cooling period were varied to establish the limits of forest community sensitivity to small and short-term temperature change. They modeled forests with 40 tree species available on fertile and poor soils with abundant water, and on both soil types with disturbances such as fires and windstorms. Baseline simulations were run for a steady-state warm climate control condition at 2854 GDD; this forest was dominated by sugar maple. Red spruce dominated the cooler control forest at 2255 GDD. Simulations were run for up to 1400 years, beginning with a clear-cut site. In the experimental runs, the initial warm-climate regime of 2854 GDD cooled in year 800 to 2255 GDD and warmed again in year 1000.

The simulated forest subjected to the 200-year cool period (a step function) underwent a change in dominant species. Sugar maple declined and red spruce increased. The increase in spruce was delayed until year 900, 100 years after the cold period began, and peaked 50 years after it ended. Then spruce declined from year 1050 on and sugar maple recovered as the simulation was run out to year 1400. A similar response occurred with American beech and balsam fir, but they were both low in abundance. The same effects were evident, but less pronounced, on the disturbed sites, where standing biomass was much less than on the undisturbed sites. When the same simulations were run for a forest on poor soils, which were dry, coarse-textured, and nutrient-poor, and with rainfall reduced to 75% of that used in the simulations described above, the effects of the climate change were somewhat obscured. The poor soils were dominated by paper birch, and white pine was of secondary importance. Red pine and trembling aspen were always in low abundance. Yellow birch showed a significant decrease in stand basal area in synchrony with the cooling, while paper birch showed no response. Again, disturbances so reduced biomass that the climate change influence was very small.

How Brief a Temperature Change Is Significant?

How brief might a climate change be and still affect forest composition? In order to test this, the same cooling event described above was used, but the duration of the cold period was changed from 200 to 100, to 50, and finally to 25 years. These simulations were run with the forest growing on good soils and with ample rainfall. The first three cold events produced significant forest changes, but when the event lasted only 25 years, no species showed a significant change in abundance. When the 100-year-long cooling was in effect from years 400 to 500, sugar maple decreased and then began to increase after year 500, and red spruce increased until year 550, but then declined. Red spruce did not reach its maximum until 150 years after the cooler period began. With the 50-year-long cooling, maximum spruce abundance occurred 125 years after the cool period began or 75 years after it ended.

How Small a Temperature Change Is Significant?

How small a temperature change might produce a significant change in forest biomass or basal area? In order to test this by forest simulation, the GDD change was reduced from the 600 GDD decrease used in the above cooling experiments to 300 GDD, to 150 GDD (equivalent to a 0.5°C mean annual temperature decrease). Sugar maple and red spruce responded significantly to the 300 GDD cooling event of 200 years duration, but no species showed a significant response to the 150 GDD cooling, even when it was extended to last 400 years. Davis and Botkin also simulated a forest response to temperature change at a hypothetical tree line. When they warmed the climate from the cold-climate limit of 747 GDD to 1197 GDD, trees began to replace treeless vegetation within 50 years. Other effects could be important. For example, the precise elevation of tree line in New England may be determined by wind or frost, and soil structure can also make a difference.

Delayed Forest Responses

Davis and Botkin (1985) commented on some of the delayed responses of the forest to climate change mentioned in the above paragraphs:

Delayed responses result from the longevity of trees and their tolerance to a wide range of climatic conditions. Mature trees have a high average survival rate even under climatic stress. Furthermore, a closed, mature tree canopy strongly influences light conditions near the forest floor, and therefore strongly influences regeneration. For example, without disturbance, a closed canopy dominated by sugar maple produced a dense shade in which relatively little regeneration took place. As the climate was made colder, spruce and fir could enter the forest only gradually. Their seedlings grew under low light conditions, where growth was slow even though the climate favored them relative to other species. Meanwhile, the cooler climate had a much smaller effect on the canopy trees, slowing their growth considerably, but decreasing their probability of survival only slightly. Under these simulated conditions, mature sugar maple trees continued to dominate the plots for a period much longer than one would predict from a consideration of the adaptations of individual species to climate. In such a case, clear-cutting a plot would hasten the change in composition from warm-adapted to cold-adapted species.

Confirmatory Evidence

Looking for confirmatory evidence for their model's predictions in the climate change–forest response record, Davis and Botkin discuss the following three climate events: (1) a cooling from 1940 to 1970 of 0.4°C in the Northern Hemisphere, which did not affect forests; (2) a warming from 1880 to 1940 of 0.6°C, which only affected tree line or second-growth forests; and (3) the Little Ice Age cooling of 2°C for more than 200 years, which produced changes in the abundances of canopy trees in forests on better soils, according to the pollen record. All three of these climate events affected biota to some degree, especially migratory birds and insects (see Lamb, 1977, and Hustich, 1952). Alpine tree line advanced upward in Sweden with the warming between 1930 and 1950 (see Woodward, 1987a). Apparently broadleaved trees are now moving northward in southern Finland, according to Erkamo (1952). Much birch dieback occurred in Nova Scotia during the 1930s and 1940s due to warmer temperatures and secondary effects, such as soil dryness and damage to mycorrhizal fungi. (This is discussed in Gates, 1980, in some detail.) In the White Mountains of California, bristlecone pine seedlings have become well established above the existing tree line since 1850. The Little Ice Age impacts on agriculture are described by Ladurie (1971).

It is interesting to note Urban and Shugart's findings for the forests of the southeastern Unites States:

Forest sensitivity to environmental stress increases with stand age, because older trees are more vulnerable to stress. This reflects the high maintenance costs of large trees, which leaves little margin for a reduction in photosynthate production. Thus, we would expect very mature trees to show the effects of climate change sooner than young trees.

Urban and Shugart (1989) go on to say that other environmental factors may contribute to forest growth, mortality, and regeneration. Thus, additional stresses would have a synergistic effect along with climate stress to produce increased local mortality. Such stresses could include nitrogen availability, on which soil moisture has a positive feedback according to Pastor and Post (1988), who modeled this interaction. They found that for northern forests on mesic soils with sufficient water, increased decomposition rates produced higher nitrogen availability and greater forest productivity. But on drier soils, drought stress reduced nitrogen availabilty and forest productivity. Another example would be the fertilizing effect of increasing atmospheric CO_2 concentrations. Carbon dioxide enrichment has been shown to improve plant water use efficiency and possibly ameliorate water stress and even nitrogen stress. (See Chapter 3 and Gates, 1985b.)

A warmer and drier climate would probably increase the frequency of fires. This might suppress some forest species and, on the other hand, open up a closed canopy to allow vigorous growth in the understory. A warmer climate might also increase pest outbreaks and fungal activity, particularly those now restricted by cold winter temperatures. These are obviously forest sensitivity factors that are coupled to climate change whose effects cannot be well estimated from our current knowledge base.

Empirical Modeling

Climate is the strongest factor influencing the distribution of vegetation throughout the world. Both climate and vegetation are collective concepts that may be defined in terms of numerous parameters (Prentice, 1990). Schemes classifying climate and vegetation together are referred to as *bioclimatic systems*. The present distributions of climate variables can be calibrated against the geographic distributions of vegetation. This empirical approach uses space as an analogue for time; in other words, it is assumed that the vegetation types occurring on the present landscape will occur on future landscapes as the climate changes, but in different locations. This approach is basically Clementsian as con-

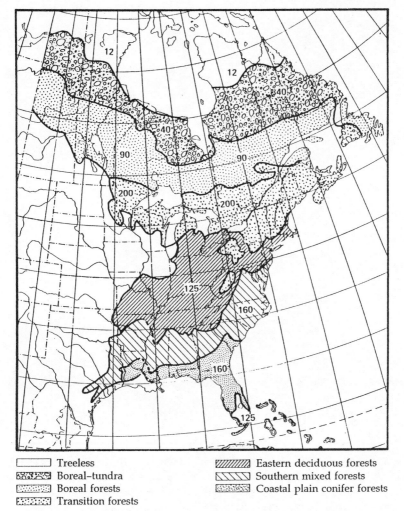

☐ Treeless	▨ Eastern deciduous forests
⬚ Boreal–tundra	◹ Southern mixed forests
⬚ Boreal forests	⬚ Coastal plain conifer forests
⬚ Transition forests	

FIGURE 8 **The present distribution of forest vegetation in eastern North America as it would be without human disturbance. Numerical estimates of average aboveground biomass in megagrams per hectare are given within each ecosystem type. (From Solomon et al., 1984.)**

trasted with the individualistic concept of vegetation dynamics—concepts discussed earlier in this chapter and in Chapter 4.

An Empirical Model of North American Forests

The present geographic distribution of North American forests is shown in Figure 8. The latitudinal forest ecosystem boundaries are reproduced

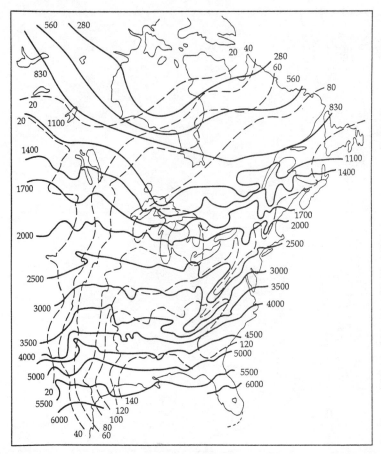

FIGURE 9   **Annual growing degree-days (GDD; solid lines) and annual precipitation (in cm; dashed lines) for eastern North America. The ecosystem boundaries of Figure 8 are well approximated by their coincidence with GDD isotherms. (From Solomon et al., 1984.)**

reasonably well by their coincidence with isotherms of growing degree-days, as shown in Figure 9. An anomaly exists near Hudson Bay because the forests farther west are limited by the low temperatures of winter as well as by the warmth of summer; a shift of GDD values for this region was made when drawing the map. Although the longitudinal ecosystem boundaries also approximate the isopleths of annual precipitation, this coincidence is not as accurate as the latitudinal match to the isotherms.

Solomon et al. (1984) took these maps and regenerated the distri-

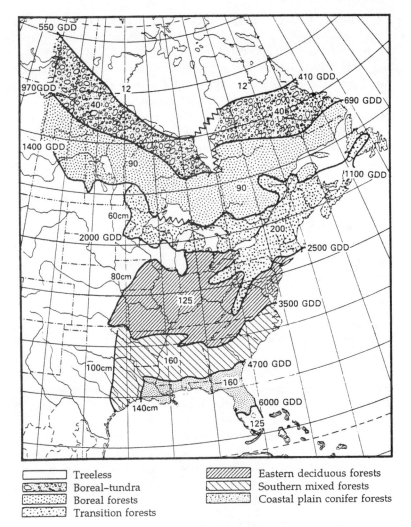

	Treeless		Eastern deciduous forests
	Boreal–tundra		Southern mixed forests
	Boreal forests		Coastal plain conifer forests
	Transition forests		

FIGURE 10   **Potential current forest vegetation as mapped from growing degree-days and annual precipitation. (From Solomon et al., 1984.)**

bution of the forest ecosystems, as shown in Figure 10, using the climate variables from Figure 9. A comparison of maps in Figures 8 and 10 shows some discrepancies, but not major ones. Solomon and colleagues then took the temperature changes predicted by GCMs under a doubling of the global atmospheric carbon dioxide concentration ($2\times CO_2$) and remapped the GDDs (maintaining the present precipitation distribution). Where the present-day GDDs were between 3500 and 6000, they added 800; between 1700 and 3000, they added 600; from 830 to 1400, they added 400; onto 560, they added 250; and, finally, onto 280 GDD, they added 50. They calculated GDD from GCM runs based on January and July temperatures, plus a sine curve to interpolate for all 12 months in the calculation. Applying the new GDDs to the forest ecosystem distributions, Solomon et al. obtained Figure 11.

Figure 11 shows that the transitional northern hardwood forests almost disappear from the United States with the temperature increase associated with $2 \times CO_2$; they are displaced into Canada. All forested ecosystems move northward, some by as much as 300 kilometers. This analysis does not imply that the forests actually *will* move, but suggests that the temperatures conducive to their growth will soon be found 300 km farther north. The area conducive to forests in eastern North America is reduced about 13%. However, there is an increase in area available for the high-carbon-density forests of the southeastern United States with the displacement of the eastern hardwoods, to yield a small net gain in standing biomass. By contrast, when the climate reaches that projected under $4 \times CO_2$, there is a large decrease in aboveground biomass in all forest ecosystems except boreal forests.

There are several difficulties associated with this empirical model. This particular model assumed that the same species assemblages growing in a given locale today will grow together somewhere in the future. In fact, at the southern border of the United States, the warmer climate will allow many subtropical plant species to grow, but the model does not allow for this; it would require human transport of these tree species, as well as other plants, from the subtropics.

Trees are slow to mature and reproduce. The expected rate of climate change is much too rapid for forest ecosystems to colonize the new regimes shown in Figure 11. The analysis does not show the transitory forests at all, but only shows where plant biomass might grow after the $2 \times CO_2$ climate change occurs.

A Climate Threshold Model for Great Lakes Forests

The ranges of four important tree species of the Great Lakes region—hemlock, American beech, yellow birch, and sugar maple—were modeled by Zabinski and Davis (1989). They used mean January temperature, mean July temperature, and annual precipitation as their defining climate variables, and used the Goddard Institute of Space Sciences (GISS) GCM output for the steady-state $2 \times CO_2$ climate change scenario described earlier in this chapter. (They also used the GFDL model, but I will not use it to illustrate possible changes since it was more extreme than the GISS model output.) The mean January temperature was used because the northern distribution limits of all four species coincide with

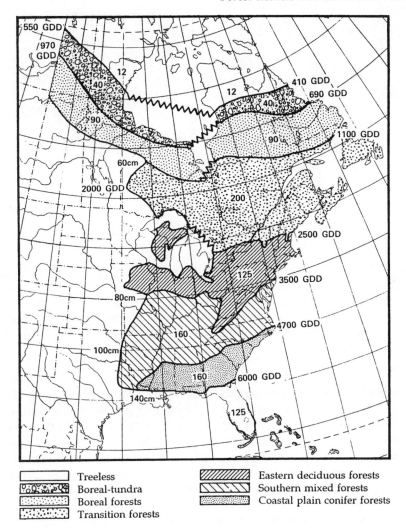

FIGURE 11 **Potential forest vegetation mapped from annual growing degree-days and annual precipitation projected from doubled atmospheric carbon dioxide. (From Solomon et al., 1984.)**

the −15°C mean January isotherm. The killing point for each of the tree species modeled is around a minimum temperature of −40°C, as described in Chapter 3. However, the GISS model output does not include temperature extremes, and hence the mean January temperature had to be used. Probably this mean temperature includes conditions where the extreme minimum experienced is about −40°C. The mean July

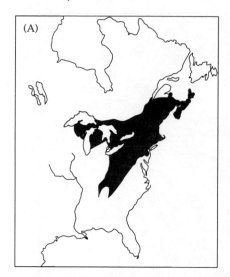
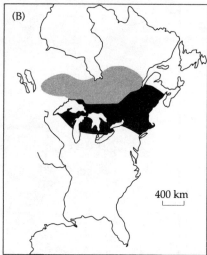

FIGURE 12   Present and future geographic range for hemlock. (A) Present range. (B) Range in 2090 A.D. under the GISS $2 \times CO_2$ scenario. The black area is the projected occupied range considering the rate of migration. The gray area is the potential projected range with climate change.(From Zabinski and Davis, 1989.)

temperature was chosen because it correlates with the southern distribution limits of yellow birch and hemlock. Precipitation has a significant effect on three of the tree species studied by Zabinski and Davis. They point out the wide variability of GCMs in projecting precipitation amounts and state that "because the global climate models do not agree on the change in precipitation, we have given a range of possible outcomes." They do not give the amounts or the ranges used.

Trees grow slowly and take many years to reach maturity. Newly established seedlings may take many years to reach a forest canopy and produce seed. Seed dispersal is quite slow, as I have indicated in this chapter and earlier ones. Tree species in the past migrated into climatically favorable areas at 10 to 40 km per century. Zabinski and Davis decided to be generous and allow the trees to move at 100 km per century. They also assumed that the change to a $2 \times CO_2$ steady-state climate would take 100 years to occur. Using the GISS GCM outputs, they drew maps of the geographic regions with climates suitable for the four species. They also showed, within the climatically favorable area, the area into which each species might migrate in 100 years. These maps are shown in Figures 12 through 15. The difference between the predicted range and the occupied range for each species is very large. This indicates that rapid dispersal of seeds and seedling establishment is critical for the distribution of these tree species by the end of the coming century if rapid climate change does indeed occur. The authors suggest that "means for artificial seed dispersal into climatically suitable regions should be considered."

Hemlock (Figure 12) prefers moist sites—often ravines if there is enough soil accumulation—and north-facing sites. Seeds germinate and grow best on moist mineral soil, moss beds, or rotting logs. Seedlings are particularly susceptible to water stress. At the northern edge of its range hemlock grows on drier sites than it does farther south. This could be a matter of competition for establishment with more mesic species, such as sugar maple, striped maple, beech, and balsam fir.

American beech (Figure 13) has a wide geographic range, from Canada to the Gulf of Mexico. Beech trees produce large, mammal- and bird-dispersed seeds. The seeds germinate readily on either mineral soil or leaf litter. Beech produces root sprouts, which gives it the ability to persist on sites where it is already established. Beech seedlings are shade tolerant and can grow well beneath hemlock. In general, beech seems to require plenty of moisture and with climate change could be limited by a drop in annual precipitation.

Yellow birch (Figure 14) has a range that is much more restricted

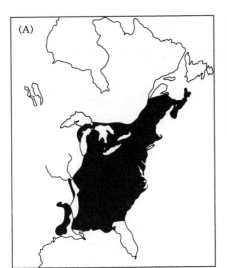 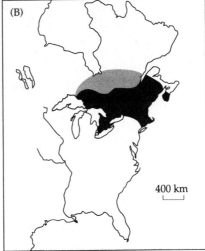

FIGURE 13 Present and future geographic range for beech. (A) Present range. (B) Range in 2090 A.D. under the GISS $2 \times CO_2$ scenario. The black area is the projected occupied range considering the rate of migration. The gray area is the potential projected range with climate change.(From Zabinski and Davis, 1989.)

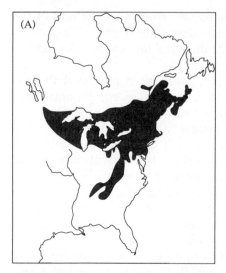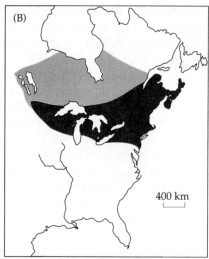

FIGURE 14   Present and future geographic range for yellow birch. (A) Present range. (B) Range in 2090 A.D. under the GISS $2 \times CO_2$ scenario. The black area is the projected occupied range considering the rate of migration. The gray area is the potential projected range with climate change.(From Zabinski and Davis, 1989.)

than that of American beech but is quite similar to that of hemlock. Yellow birch seeds, like hemlock seeds, are very small and germinate readily on moist mineral soil. Apparently seedlings can tolerate hot, dry conditions once their roots are well established. They do well with plenty of light. Yellow birch has the largest potential climate range of the four tree species because of its ability to live on drier sites.

Sugar maple (Figure 15) produces very large seed crops annually. These seeds grow particularly well in leaf litter on the forest floor under low light levels, but need adequate moisture. They form large beds of seedlings that hang on for several years, but then experience high mortality if a light opening in the canopy does not occur. Sugar maple is a very competitive species because of its ability to survive in the shade and to grow rapidly when it does receive light. Because of its large seed cohorts, sugar maple is expected to do well after climate warming. Zabinski and Davis write, "Because of the high number of seedlings produced, and the high amounts of mortality necessary to reduce the population as the cohort ages, there is potential for sugar maple to produce a cohort that is adapted to the conditions at the time of establishment." Sugar maple has a greater tolerance for low moisture levels than the other three species described here, but it cannot, for example,

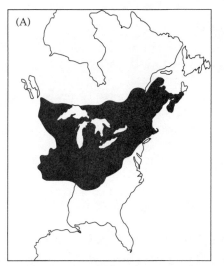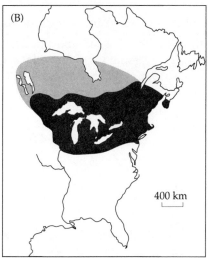

FIGURE 15   Present and future geographic range for sugar maple. (A) Present range. (B) Range in 2090 A.D. under the GISS $2 \times CO_2$ scenario. The black area is the projected occupied range considering the rate of migration. The gray area is the potential projected range with climate change.(From Zabinski and Davis, 1989.)

grow at the edge of the prairie. With the warming climate, sugar maple is expected to continue to grow in the Great Lakes region, but it may be restricted to fewer sites, particularly if the climate becomes drier.

Zabinski and Davis suggest that under the GISS $2 \times CO_2$ scenario, many hardwood species in the western and southern parts of the Great Lakes region will be endangered. The marginal climate will stress the trees and produce increased mortality due to insects and pathogens. Successional species, such as trembling aspen and paper birch, will increase; these will provide fodder, and the deer population will increase. (I believe it is likely to be too warm and perhaps too dry for paper birch to thrive.) Since beech will be greatly reduced, its seeds will not be available to the many bird and small mammal species that now depend on them for food. Fires may become more frequent in these forests with the climate warming, although Zabinski and Davis suggest that the GISS scenario indicates a wetter situation that may tend to dampen fire frequency. Artificial methods may need to be developed for broadcasting seeds or seedlings ahead of the northward-moving climate front in order for all four tree species to colonize more rapidly than they would naturally. For sugar maple, it may be necessary to select warm-adapted seeds from the southern edge of its range.

Future Forests from Response Surfaces

In Chapter 4 I discussed the derivation of response surfaces for seven forest tree genera and for sedges and prairie forbs as derived by Bartlein et al. (1986). Webb (1986) reconstructed the climate of the past 18,000 years from the pollen record in eastern North America. Overpeck and Bartlein (1989) then used these transfer functions to project the future distributions of the seven forest trees, sedges, and prairie forbs based on the GISS, GFDL, and OSU $2\times CO_2$ greenhouse warming models. Each of the climate scenarios produced similar patterns of vegetation change, with some exceptions.

Using these results, Overpeck and Bartlein projected that with the $2\times CO_2$ climate warming, the southern range limit of spruce may move northward by several hundred kilometers. Spruce and the northern pines will decrease in abundance in the eastern United States. The transition forests will become increasingly deciduous. There will likely be large increases in oak abundance in the northern Great Lakes and New England areas. Oak and the northern pines may move northward by as much as 500 km. The models all showed an expansion of prairie forbs throughout the eastern United States, thus suggesting that forest biomass may decline throughout the region. This was primarily the result of higher temperatures, since it occured in the wetter-than-present GFDL and OSU scenarios as well as in the drier GISS scenario. The northern boundary of the southern pines may expand northeastward by as much as 500 to 1000 km. Birch will be displaced northward a great deal. In eastern Canada, the boreal forest will become more pine-rich than at present. The Canadian tundra will shrink as trees become established in it.

Overpeck et al. (1991) reported this sequence again, but with some added information concerning rates of change in forest composition. They emphasized the individualistic concept of community structure, with each species moving according to its own requirements. They found that changes in the equilibrium distribution of natural vegetation in North America over the next 200 to 500 years could be larger than the changes that occurred during the past 7000 to 10,000 years. This rate of change would be comparable to the change that took place over the 1000- to 3000-year period of most rapid deglaciation.

Bioclimatic Models of Vegetation

Many global bioclimatic classification schemes that associate vegetation types and climate regimes have been devised, including those of Grisebach (1838), von Humboldt (1867), Köppen (1900, 1936), Thornthwaite (1931, 1933, 1948) Holdridge (1947, 1949), Troll and Paffen (1964), and

Box (1978). Prentice (1990) decided to test these bioclimatic models by using four of them (Holdridge, Thornthwaite, Köppen, and Troll and Paffen) and two gridded global climate data sets to simulate the global distribution of vegetation. These tests were devised not only to demonstrate the accuracy with which global vegetation can be simulated, but also to find out the sensitivity of vegetation to climate and to how climate is defined. Prentice discovered many difficulties inherent in each of the bioclimatic classification schemes. Part of the problem is that the climate regimes of the world are very coarsely defined. As Prentice states, "The vegetation types must be 'vegetatively meaningful' to be useful in a study of biosphere–atmosphere exchange. The vegetation types must also be 'climatically meaningful' to be simulated from climate."

Prentice found that "only 38 to 40% of the observed land surface mapped as thirty-one vegetation types, could be replicated by applying the four schemes to two global climate data sets." She then had to make some modifications of the gridded climates and regroup some of the vegetation types. The climate regimes were more coarsely gridded than were the vegetation regimes. She also had to refine the physiognomy of the vegetation types in order to achieve a classification more descriptive of the canopy features that influence atmosphere–biosphere gas and energy exchanges. She found that the bioclimatic schemes of Thornthwaite, Köppen, and Troll and Paffen, which were seasonally based, were difficult to simulate using seasonal precipitation. The Holdridge classification scheme was based on the more reliable and easier to simulate annual climates. Finally, with these alterations, 77% of the predicted vegetative landscape corresponded with the present distribution of vegetation.

Predicting Global Warming Effects on Biomes

Prentice's results (Prentice, 1990) determined that the Holdridge scheme can be used to predict changes in vegetation distribution that are likely to occur under global warming. This complex scheme, shown in Figure 16, classifies climate according to the average total annual precipitation (APPT) and the annual potential evapotranspiration ratio (APETR), which is the quotient of the annual potential evapotranspiration and the average annual precipitation. Then a triangular diagram is formed, with APPT as two of the axes and APETR as the third axis. All axes are logarithmic. A mean annual biotemperature is defined as the average over the year of daily, weekly, or monthly temperatures, and is given the value 0 if the temperature is less than or equal to 0°C. The mean annual biotemperature axis is placed on the diagram perpendicular to the base of the triangle. The diamonds formed in this diagram are

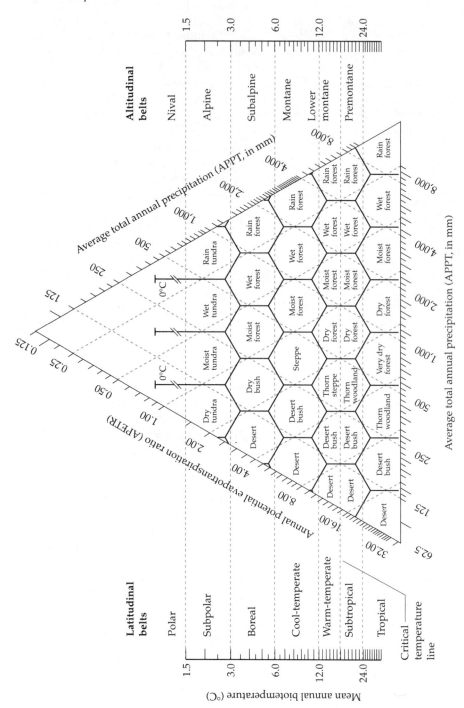

◄ FIGURE 16   The Holdridge Life Zone classification, shown here as a "Holdridge triangle." The mean annual potential evapotranspiration ratio forms one side of the triangle. The average total annual precipitation forms two sides of the triangle. The mean annual biotemperature is superimposed horizontally on the triangle. Each hexagon defines a bioclimate named for a vegetation assemblage. (From Prentice, 1990.)

defined climates. The hexagons form the **Holdridge Life Zones** and indicate vegetation assemblages; these are identified in Figure 16. There is a further line of demarcation in the Holdridge diagram, along a critical temperature limit based on the occurrence of killing frost, which separates warm temperate from subtropical zones.

The following are Prentice's observations using her revised Holdridge scheme to project changes in various vegetation communities under the GISS $2 \times CO_2$ scenario. She does not explicitly state the warming scenario but makes qualitative statements.

Tropical/subtropical seasonally humid bioclimates   All vegetation biomes with a mean annual biotemperature above 17°C and an APETR of less than 2 are in this category. This includes tropical evergreen rain forests, temperate evergreen seasonal broadleaved forests, drought-deciduous woodlands (which include woody plants but not trees) and drought-deciduous forests. These vegetation communities are located across much of South America, Africa, and Asia south of 10° north latitude. They exhibit very little climatic seasonality. With the projected warming under $2 \times CO_2$, the fate of tropical rain forests will depend on precipitation. If precipitation were to remain the same as today or decrease, APETR would increase and what is now the tropical rain forest belt would evolve into a more arid and warmer forest type, such as drought-deciduous. If precipitation were to increase such that APETR is the same or less than at present, the tropical rain forests might become what we would imagine as "extra-equatorial rain forests"—a type that does not exist today.

Temperate moist bioclimates   Vegetation communities with a biotemperature greater than 6°C and less than 17°C and an APETR of less than 2 make up the temperate moist bioclimates. This includes broadleaved cold-deciduous forests, evergreen broadleaved woodlands, evergreen needle-leaved woodlands, tall grasslands, and short grasslands. These vegetation types are largely located in North America north of Mexico, across Europe and Asia north of India, and at the southern tip of South America. They are characterized by great climatic seasonality. With global warming associated with $2\times CO_2$, one would expect from the Holdridge classification that grasslands would shift toward a drier warm-temperate thorn-steppe vegetation or shrubland if there is no increase in precipitation, or toward a broadleaved evergreen woodland if precipitation increases sufficiently to keep APETR at about 1.14. Prentice makes the following important point: "The dominance of grass versus trees in cool wet regions depends on the balance of the availability of near-surface water to sustain the more shallow-rooted grasses and the availability of deeper water to support the deeper-rooted trees. Such a specific and complex water balance exceeds the capacity of the simple bioclimatic scheme used here."

Arid bioclimates The arid bioclimates are generally found where the APETR is greater than 2. Arid conditions may occur within all biotemperature regimes from boreal through tropical, as seen in the lower left part of the Holdridge triangle (Figure 16). This bioclimate is large because it is a "catch-all" for many arid types that are climatically inseparable. Making climatic distinctions among arid types of vegetation is difficult because of seasonal variation in precipitation and temperature. In the Mojave Desert, for example, production of organic matter increased by approximately 250% from a dry year to a wet year. Prentice emphasizes the difficulty of separating arid and semiarid vegetation climatically because of the importance of soil characteristics. When precipitation drops below 500 mm per year in the tropics, the soil type becomes increasingly important in determining vegetation communities. For example, deep sandy soils support savanna, while stony soils support thornbush. Under even drier conditions, savanna becomes grassland and thornbush becomes a shrub semidesert. Global warming with no increase in precipitation would cause further drying of these arid regions, and they would begin to acquire desert vegetation. If precipitation increases with the warming, the arid grasslands and shrublands may become transition zones to drought-deciduous forests and tall grasslands.

Boreal–polar bioclimates Boreal–polar bioclimates have biotemperature less than or equal to 6°C. Vegetation communities here include temperate/subpolar evergreen needle-leaved forests, cold-deciduous needle-leaved forests, cold-deciduous needle-leaved woodlands, tundra, and ice/polar desert. Excluded from this region are broadleaved evergreen species whose leaves and buds cannot survive temperatures below −15°C. Temperature limits of −15°C to −40°C restrict many of the broadleaved temperate forest species found to the south of this region and allow needle-leaved forests to dominate (see Chapter 3 for further discussion of this topic). Prentice refers to Woodward (1987a) and states, "Although APETR is not the equivalent of heat sums or minimum temperatures, it seems to serve here reasonably well as an analogous measure." The expected warming under $2 \times CO_2$ will be greatest in the high latitudes (with the exceptions described in Chapter 1.) If this warming occurs, temperate vegetation will invade boreal regions while tundra will be displaced by boreal vegetation types.

Another Holdridge Life Zone Example

Emanuel et al. (1985a) also used the Holdridge Life Zone bioclimatic classification scheme to anticipate the effects of global warming on terrestrial ecosystems. In general, these investigators obtained results in agreement with Prentice's except at high latitudes, where they improperly ignored the enormous seasonality of the climate and did not apply the biotemperature scheme properly. [This was pointed out by Rowntree (1985) and subsequently corrected by Emanuel et al. (1985b).] Emanuel et al. found that under $2 \times CO_2$ global warming, global forest area will decrease from 58.3% to 53.9%; grasslands will increase from 17.6% to 21.6%; deserts will increase from 20.6% to 21.5%; and tundra will decrease from 3.5% to 3.0% of the global land area, excluding ice-covered areas. Tropical forests will increase from 25% to comprise 35% of the forested areas; subtropical forests will decrease from 16% to 13%; warm-temperate forests will stay unchanged at 21%; cool-temperate forests will increase from 15% to 16%; and boreal forests will decrease substantially, from 23% to 15%. "Boreal moist forest is replaced by cool temperate steppe or, in limited areas, by cool temperate forest or by boreal dry bush. Boreal wet forest is replaced by cool temperate forest or boreal moist forest."

Emanuel et al. pointed out that only temperature change was used in the projected climate scenario and that precipitation changes could make a substantial difference. They also reiterated the difficulty of working with climate models whose spatial resolution is much lower than the resolution of vegetation communities.

An Alternative Analysis of the Canadian Boreal Forest

Sargent (1988) criticized the Holdridge scheme for not being based on a sufficient number of climate variables. He used a model developed by Box (1981), based on eight climate parameters controlling plant survival, to model the response of the Canadian boreal forest to climate change. The eight parameters used were: minimum and maximum monthly average temperature and precipitation (TMIN, TMAX, PMIN, PMAX); annual range of monthly average temperatures (DTY); annual precipitation (PRCP); average precipitation of the warmest month

(PMTMAX); and a ratio of the annual precipitation to the Thornthwaite potential evapotranspiration for the year (MI). In Box's model, if at a given location the value of any one of these eight climate parameters fell outside its allowable limits for a given species or group of species, then the climate was considered unsuitable for their survival. Sargent gives lower limits for each of the eight climate parameters for the Canadian boreal (spruce and fir) forest, and upper limits for three of them (he could not establish upper limits for the last five parameters listed below). Sargent's southern and northern limits are: TMAX, 20, 13; TMIN, −8, −29; DTY, 60, 10; PRCP, 100; MI, 0.6; PMAX, 25; PMIN, 5; and PMTMAX, 55. (All units are in °C and mm except MI, which is dimensionless.)

Sargent used the GFDL and GISS $2 \times CO_2$ GCM climate scenarios for Canada and applied the projected climates to the boreal forest requirements. A characteristic feature of both models was the warm arctic winters they projected; warming was much less pronounced in summer. Eastern Canada warmed two to three times as much as western Canada. Winter temperatures increased 6°C to 10°C in the east and only 2°C in the west for the $2 \times CO_2$ climate. The boreal forest shifted northward by about 5 degrees and was greatly reduced in area.

Boreal Forest Change with GDD

The boreal biome occupies a large fraction of the Northern Hemisphere at high latitudes in a band of more than 1000 kilometers from north to south across the Eurasian and North American continents. This represents about 23% of the forested area of the world and about 14% of the total forest biomass. The boreal biome is dominated by coniferous forests. The boreal region is of particular significance from the standpoint of climate change, since it is at high latitudes where the greatest warming is expected to occur. (There might be some exception to this produced by the oceans, as suggested in Chapter 1.) Not only will greenhouse warming affect the growth of, or the even presence of, the boreal forests, but also their soils, which hold between 16% and 24% of the global soil carbon. Increasing temperatures might result in the release of considerable CO_2, thus exacerbating the greenhouse effect.

Temperature is a very important determinant of growth for boreal forests, where it is usually more limiting than moisture. A cool climate during the growing season limits the productivity of boreal forests. Kauppi and Posch (1985, 1988) modeled the growth response of the boreal coniferous forest in Finland to temperature. All of Finland is within the boreal region today. The mean annual temperature is 4.5°C at its southern edge and −1.0°C at the northern border. The country is fairly uniform with respect to soil conditions and topography. The objective of Kauppi and Posch's study was to understand the sensitivity of boreal forests to climate warming. In particular, they wished to learn whether the most sensitive forests would be in the north or the south, or in the more continental or more maritime regions.* For their work in Finland they used GDDs calculated from average daily temperature sums above a threshold of 5°C. Each of 19 sites in Finland was assigned a GDD sum and a value of continentality. The most maritime region had a continentality of 8°C, while the most continental site had 13°C.

The Finns have an enormous data base of growth measurements of their forests. The mean annual tree growth rate is approximately a logistic function of GDD sums. Its shape takes into account the drought limitation of growth under high GDD conditions. Kauppi and Posch assumed a uniform 2°C increase from all baseline temperatures, both summer and winter, and calculated new GDD sums. This calculation was site-specific and included the degree of continentality at each site. Then the average annual tree growth rate was determined using the logistic regression with the new GDD sums. The results showed that a given climate warming would yield the greatest absolute increase in tree growth in warm (southern) and maritime parts of the Finnish boreal biome. In terms of relative or percentage increase in growth rate, the sensitivity increased northward and toward maritime areas.

In their second paper, Kauppi and Posch extended this analysis to the entire circumpolar boreal zone. They used the GISS $1 \times CO_2$ and $2 \times CO_2$ GCM outputs for climate warming over a grid covering the Northern Hemisphere between 38° and 70° north. The grid size was 5 degrees in the east–west and 4 degrees in the north–south direction. Observed mean monthly temperatures were converted into GDD sums for each grid box using a threshold temperature of 5°C and an assumption of continentality of 3°C.

The northern and southern boundaries of the boreal region were approximately defined by the 600 and 1300 GDD isopleths. The isopleth positions under the GISS $2 \times CO_2$ scenario were then calculated. Under this scenario the boreal forest zone shifted northward by 500 to 1000 km (Figure 17). The boreal zone shifted nearly off the continent of Europe with a steady-state warming under $2 \times CO_2$.

Kauppi and Posch applied the GDD sums for $2 \times CO_2$ steady-state climate to the logistic growth function for the boreal forests of the Northern Hemisphere, although the forest data used was only from Finland. Their results showed that the greatest growth of boreal forest

*The more continental climate, with the largest variability, is toward the north. A measure of continentality is the standard deviation of the monthly temperature frequency distribution: the broader the distribution, the greater the continentality and the difference between summer and winter temperatures.

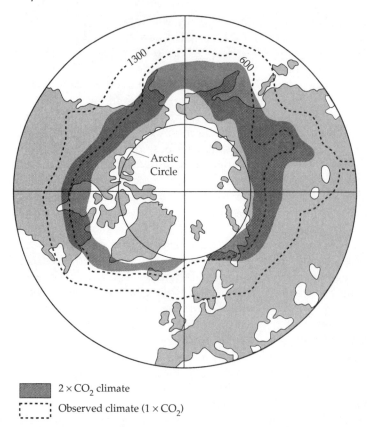

[■] $2 \times CO_2$ climate

[⌐ - - - ⌐] Observed climate ($1 \times CO_2$)

FIGURE 17   **The northern and southern boundaries of the boreal forest are approximately defined by the 600 and 1300 growing degree-day isopleths. These are shown in their current positions and in the positions they would occupy under a 2×CO₂ climate warming. (From Kauppi and Posch, 1988.)**

trees would occur over northern maritime regions. Northern continental regions were too cold to maintain much productivity, even under the $2 \times CO_2$ climate. In southern parts of the boreal zone, it was assumed that moisture stress restricted growth. We know that tree growth does not respond at all to a large increase in winter temperatures since they are still well below the threshold for growth, but tree growth does respond to a modest increase—from say 3°C to 7°C—in growing season temperatures, where the threshold for growth is 5°C. These results suggest to Kauppi and Posch that

. . .despite the counteracting effect of relatively lower estimated climatic warming over maritime regions, maritime ecosystems might nevertheless respond more strongly than continental ecosystems. Highest responses were estimated in Labrador, southern Greenland, Iceland, northern Fennoscandia and around the Bering Strait.

A basic assumption of all of these calculations is that boreal forests are in equilibrium with the present climate and would be in equilibrium with the $2\times CO_2$ climate. The results are valid if the greenhouse gases stabilize at the $2\times CO_2$ level and the forests are given time to become adapted to the new climate.

Forests and Other Organisms

This and the preceding chapters have described how individual plant species may respond to climate change, how forest communities slowly undergo succession, how the forest biomes have been altered in the past, and how they may be affected in the future by global warming. Given the vast changes that could occur in the forest landscape, how will the animals who live in that landscape be affected? The next chapter addresses this and other questions.

CHAPTER SIX

Ecosystems

IN PREVIOUS CHAPTERS I described how individual plant species may respond to climate change and how forest communities may slowly undergo succession or be altered in various ways. I have not discussed the animal component of ecosystems, nor ecosystems in general. But first, what is an ecosystem?

Ecosystems

An **ecosystem** is a subdivision of the landscape, a geographical area that is relatively homogeneous and reasonably distinct from adjacent areas. An ecosystem is made up of three components: organisms, environmental factors, and ecological processes. An ecosystem may be a bog, sand dune, meadow, grassland, lake, stream, or forest. Organisms are represented by populations of individual plant and animal species which taken all together make up communities. Environmental factors include temperature, moisture, wind, radiation, atmospheric chemistry, soil texture and chemistry, and events such as hurricanes, tornados, floods, frosts, droughts, and fires. Ecological processes include energy flow, nutrient cycling, productivity, plant and animal competition, predation, and other organismic interactions.

The environment is always changing, sometimes quickly and sometimes slowly. The chemistry of the atmosphere changes and with it the patterns of transfer of energy and mass, including changes in temperature, moisture, pressure, and winds. These changes affect the hydrosphere, geosphere, and biosphere. The consequence of an environmental change on an ecosystem will depend on both the magnitude of the change and its duration. Ecosystems are resilient and reasonably stable, but there is a stress level beyond which they will not return to their former state. Species respond individually to climate changes and to other factors such as acid rain. Organisms usually lag in their responses

162

to change, lags that may be days, weeks, months, or years. Ecosystems may take decades or even centuries to respond to a change. Individual organism responses to climate will change the composition of plant and animal communities through time, which in turn will feed back to the atmosphere and soil.

Primary Productivity and Feedbacks in Ecosystems

The first direct effect of climate and atmospheric changes on ecosystems is on their **primary productivity**—the amount of biomass the plants produce by photosynthesis. Changes in primary productivity then cause shifts in plant–plant competition and responses among associated populations of animals, as well as strictly physical feedbacks through energy flow and nutrient cycling. Many plants and animals will respond independently of the ecosystem as a whole, and the ecosystem will slowly change its form. If the albedo of an ecosystem changes, this feeds back directly on the atmosphere (see the discussion of desertification in Chapter 7). Higher temperatures will increase an ecosystem's respiration rate and this will release additional CO_2 to the atmosphere, which will in turn increase temperatures through the greenhouse effect. Drier conditions under certain circumstances will reduce plant productivity, decrease vegetation cover, increase soil erosion, increase loss of nutrients, and further reduce plant growth. On the other hand, increased atmospheric CO_2 levels will enhance photosynthesis and productivity, increase carbon allocation to roots, increase mycorrhizal activity, increase nitrogen fixation, and result in more plant growth.

Plant Metabolism and Climate

Plants are grouped into three main classes according to their metabolic pathway for the assimilation of CO_2. If a three-carbon acid is produced, the plants are classified as C_3 plants; if a four-carbon acid, they are C_4 plants; and finally a group of plants that store CO_2 at night and assimilate it in daylight are referred to as CAM plants. CAM plants are generally succulents growing in arid regions. C_4 plants, such as corn, are better adapted to higher temperatures than are C_3 plants, such as wheat; but C_3 plants usually respond to increased levels of atmospheric CO_2 with greater increases in photosynthetic rates than do C_4 plants. Among natural ecosystems, where there is a mixture of C_3 and C_4 plants, one can expect the balance to shift with warmer temperatures and increasing levels of atmospheric CO_2. This will be particularly true of grasses where one would find an increase in the ratio of C_4 to C_3 grasses in some places. In drier regions one would expect an increase in the proportion of CAM plants.

In case the reader assumes that these three simple classes allow us to exactly predict the shift in plant species composition within a given ecosystem, they should be reminded that some species are troublesome and do not perform as expected; see Bazzaz (1990) for a review of this subject.

Ecosystem Responses to Climate

The responses of one ecosystem to rising temperatures, increasing or decreasing precipitation, increasing CO_2 concentration, acid deposition, or other environmental impacts may be quite different from that of another ecosystem. What happens within an arctic tundra will be quite different than what happens within a Mediterranean chaparral ecosystem, a boreal forest, or a tropical rain forest. According to Walker (1991), "In general the changes will be most marked in those ecosystems where the structure and composition are strongly influenced by limiting conditions of temperature or rainfall." Climate projections for Europe using GCMs indicate an extension of the normal Mediterranean summer dryness, despite an increase in winter precipitation. The longer summer drought will likely shift the species composition and will certainly lead to reduced plant cover, which will probably further reduce soil moisture availability.

Walker points out in his review of the ecological consequences of climate change that "vegetation will change more as a result of rare and extreme events than in response to a change in average conditions." He gives an example from Austin and Williams (1988) of vegetation changes in Australia that are the result of grazing animals and El Niño/ Southern Oscillation (ENSO) events (see Chapter 7 for a description of ENSO). The scrub coniferous tree *Callitris glaucophylla* developed on open grassland as a result of the rains that followed the severe ENSO event of 1876–1878, in combination with heavy grazing by animals and fire suppression by humans. Numerous other examples of this type may be found in the American West, where shrubs have taken over grasslands as the result of overgrazing and drought combined. For example, no shrubs were found on 60% of the 58,000-hectare desert grassland of the Jornada Experimental Range in southern New Mexico when it was surveyed in 1858. Only 5% of the area supported mesquite and the other 35% supported scattered shrubs. By 1963, 73% of the area was dominated by dense stands of mesquite, creosote bush, or tarbush, according to Buffington and Herbel (1965). Scientists studying the response of plants to higher levels of CO_2 are finding that woody plants, like mesquite, grow larger and faster than warm-season forage grasses, and therefore may outcompete them.

Hunt et al. (1991) simulated the responses of temperate grasslands to climate change using a grassland ecosystem model that treated the seasonal dynamics of shoots, roots, soil water, mycorrhizal fungi, microbes, soil organisms, nitrogen levels, and soil organic matter. Forty-year simulations were run for several climate change scenarios. Precipitation and CO_2 concentration produced most of the responses among plants, animals, and microbes. Elevated temperature extended the growing season but reduced photosynthesis during the summer, with little net effect on annual productivity. Doubling the atmospheric CO_2 concentration caused increases in primary production, even with nitrogen limitation, and led to greater storage of carbon in plant residues and soil organic matter.

In Chapter 3 I discussed the effects of winter or spring warmings on the frost hardiness of northern hardwood trees—particularly of their roots, which are especially sensitive to sudden cold following unseasonable warming. If greenhouse warming results in more of these types of climate excursions, one would expect considerable tree mortality to follow. The frequency of extreme heat events during the summer will also increase with greenhouse warming, and this will have a considerable impact on old and young individuals in a community (Mearns et al. 1984).

The seasonality of precipitation is critical to most ecosystems. A change in the seasonality may produce quick changes in the vegetation composition and an associated shift in animal populations. A shift in the pattern of rains on the Serengeti Plain of Africa, for example, could result in a quick response by the wildebeest herds, which would move to nearby regions; or a complete drying of water holes, which would result in the massive die-off of large mammals.

Loss of Species Diversity

There is great concern among ecologists over the loss of species diversity in the world as habitats shrink or are lost altogether. There is no question that large extinctions have occurred in the past, some probably the result of climate change, many the consequence of human activities, and others the result of cataclysmic disturbances. It is almost certain that as the climate warms and as extreme events become more frequent, species diversity will diminish. The greatest species diversity today is in tropical regions with equable climates, and those species are the least tolerant of large temperature and moisture variations. But how can one predict exactly which species will be most vulnerable to climate change during the coming century?

Peters (1992) suggests that those species or populations most susceptible to climate change will fall into one or more of the following classifications.

1. Peripheral populations of plants or animals that are at the contracting edge of a species range.
2. Geographically localized species. Many currently endangered species exist in extremely limited habitats.
3. Highly specialized species. Many species have, for example, a close association with only one other species, like the snail kite that feeds exclusively on the Pomacea snail in Florida wetlands.
4. Poor dispersers. Many trees have heavy seeds which may not disperse far. Some tropical forest birds will not cross over even a small barren patch of land to another piece of forest.
5. Montane and alpine communities. Populations of plants and animals on mountains may literally be pushed off the mountaintops as the climate warms.
6. Arctic communities. Climate warming is expected to be greatest at high latitudes. Therefore, organisms of these regions may be dealing with the most rapid change.
7. Coastal communities. Because sea levels are expected to rise considerably, many shoreline communities will be inundated.

Those animal species that are most mobile are likely to find survival easiest as rapid climate change takes place, provided their food source is still available. Trees are probably the slowest moving of all organisms. But certain birds, mammals, insects, and pathogens are so tightly coupled to specific forest trees that they may not move and survive beyond the boundary of those species. In Chapter 5 I mentioned how the Kirtland's warbler only breeds in and nests on the ground under intermediate-age jack pines growing on well-drained sandy soil in northern Michigan. As the climate warms, the jack pines may move north into less well-drained soils, and warbler nests could no longer be successful on the moister soils. Some scientists believe the Kirtland's warbler may be the first species casualty of greenhouse warming.

Birds

The nesting requirements of the Kirtland's warbler are particularly severe, but other bird species are also vulnerable. Birds may fly to a more tolerable climate as conditions change, but unless their food source can be found there it will do them no good. In addition, migratory birds must have suitable "stop-over" habitats along their migration routes and must find adequate food there as well. If these habitats are destroyed by climate warming, migration to summer or winter ranges will not be possible.

Winter Isotherms and Bird Distributions

What determines the geographical distribution of a bird species? Is it food abundance, climate, habitat structure, competition and predation, physiological characteristics (such as metabolism or morphology), or other factors? To find out, Root (1988a,b) studied the distributional boundaries of wintering North American birds. She drew distribution and abundance maps for 148 species of land birds wintering in the contiguous United States and southern Canada and compared these with maps of six environmental factors: (1) average minimum January temperature, (2) mean length of frost-free period, (3) potential vegetation, (4) mean annual precipitation, (5) average general humidity, and (6) elevation. The comparisons showed that isograms for the average minimum January temperature, mean length of frost-free period, and potential vegetation correlated with the northern range limits of 60.2%, 50.4%, and 63.7%, respectively, of the wintering bird species. Only two environmental factors—potential vegetation and mean annual precipitation—were found to coincide with the eastern range boundaries, for 62.8% and 39.7% of the species, respectively. At the western boundaries, mean annual precipitation distribution coincided for 36.0% of the species, potential vegetation for 46.0%, and elevation for 39.7%. Remember that mean isotherms generally run across the United States along a southwest-to-northeast trend, and mean isopleths of precipitation run along a south-to-north trend that parallels, for example, the forest grassland border (see Figure 9 in Chapter 5). Figure 1 shows the northern winter range limit for the eastern phoebe and the $-4°C$ average January minimum temperature isotherm.

Root (1988b) questioned why the northern boundary for so many wintering bird species coincided with the average minimum January temperature. She calculated the metabolic rate at the northern boundary for 14 of the 51 songbird species whose northern winter range limit related to a particular average minimum January temperature and whose basal metabolic rates had been measured. Within a certain temperature range, called the **thermal neutral zone**, a bird's resting metabolic rates does not change from the basal rate with ambient temperature. Below this range, the resting metabolic rate increases linearly with decreasing temperature to supply the energy necessary to balance heat loss. (See Gates, 1980, for a description of the energy budgets of animals.) The mean metabolic rate at the northern boundary for each species Root studied turned out to be 2.49 times its basal metabolic rate, with a standard error of only 0.07.

FIGURE 1   The distribution and abundance of the winter range of the eastern phoebe. The northern boundary lies very close to the −4°C isotherm of January minimum temperature (heavy solid line). The contour lines are at 20%, 40%, 60%, and 80% of the maximum abundance. The tick marks point toward lower abundance. (After Root, 1988a.)

Root next calculated the basal metabolic rates of 36 of the other 37 songbird species, whose basal rates had not been measured but could be estimated from the dependence of metabolic rate on mean body weight. When she determined the ratio of the northern winter boundary metabolic rate to the basal metabolic rate for these birds, she also got 2.5. So the "2.5 rule" seems to be remarkably consistent for songbird species, whose weights range from 5 grams (wrens) to 448 grams (crows), diets from seeds to insects, and winter northern borders from Florida to Canada.

Greenhouse climate warming may move these mean January minimum isotherms northward by as much as 250 kilometers during the coming century, particularly since winter temperatures are expected to increase more rapidly than summer temperatures. Many of the bird species studied by Root seem to be able to follow this movement of isotherms in synchrony with them. For example, Root has found that some birds will respond to the previous year's temperature; some bird species were found farther north during a year following a warm year. But for some of the bird species, their habitat or food requirements may not permit them to shift their winter ranges at the same rate as the climate change. The result will be a reorganization and restructuring of the bird communities as they are known today. If precipitation changes occur in winter months, then the western range limit for some of the songbirds will shift, most likely in response to the movement of plant food sources.

Bird Migration and Climate Change

Many sand-nesting bird species, including many pelagic birds such as seagulls, will have their nesting habitats disrupted (Figure 2A). Arctic-breeding shorebirds nest on the flat, open tundra among mosses, lichens, and low-growing plants (Figure 2B). If greenhouse warming changes the plant structure of the tundra and taller-growing shrubs move in, will these shorebirds be able to adapt? It is unlikely they will. Many of these same arctic-breeding shorebirds stop along Delaware Bay during their northward migration to feed on the eggs of horseshoe crabs and renew their energy supply. If the climate warms, the crabs may start spawning before the birds arrive. The nesting success of the arctic-breeding shorebirds is tied to insect hatches in the far north. If the tundra warms weeks earlier than it does now, the insect hatches might occur before the birds arrive, and the birds might be unable to obtain sufficient food and achieve enough weight gain for their southern migration in the autumn. This is an example of the importance of timing in ecosystems and how a shift for one animal species could be disaster for another. We do not have a very thorough understanding of these interactions and exactly what their consequences may be. It is entirely possible that bird populations will be more adaptive than we give them credit for—but then again, they may not be. (See Peters and Lovejoy, 1992, for further details concerning the response of animals to climate change.)

The annual northward migration of the lesser snow goose follows the melting snow line. When the snow melt is delayed, the population of geese builds up south of the snow edge. If global warming produces greater precipitation and increased snow depth in the far north, and along with increased cloudiness, there could be a delay in snow melt. The geese would remain south of the snow line, and the delay could cause heavy, destructive feeding on vegetation there. Normally, when the geese clip the leaves of grass and sedges, the plant parts regenerate rapidly. In spring, before the above-ground growth of vegetation begins, the geese grub for roots and rhizomes. According to R. Jefferies of the University of Toronto, as reported by Miller (1989), a grubbing goose can completely clear one square meter of land in an hour and leave it a mudflat. Thousands upon thousands of snow geese may leave productive salt marshes a vast mudflat. Not only will there be destruction of salt marshes, but in nearby freshwater ponds the geese will pull up sedge shoots and eat the basal parts. During successive years there will be less and less vegetation available as food for the geese. Already, during four successive springs the amount of vegetation remaining to support geese at La Perouse Bay, Ontario has decreased between 8%

FIGURE 2   Bird nesting habitats may be disrupted by climate warming. (A) A seagull nest built on the sands of Baja California. (B) A nesting Arctic tern in Alaska; a warmer climate could change vegetation patterns and disrupt the tern's nesting habitat. On the other hand, the population of chinstrap penguins (C) appears to have been favored over the past decade by warmer winters that meant increased breeding success. (A, photo by Gary J. James; B, photo by Robert Stottlemyer; C, photo by P. R. Ehrlich, Stanford University. Courtesy Biological Photo Service.)

and 12%, and the weight of goslings at the end of summer has fallen nearly 12% in a decade.

Antarctic Seabirds and Climate

In the Antarctic, chinstrap penguin populations more than tripled between the 1940s and the 1980s. Chinstrap penguins (Figure 2C) overwinter in the open sea, and their breeding success rises abruptly after warmer winters. Is this increase a response to global warming? Likewise for the skua, which feeds largely on minnowlike antarctic silverfish 2 to 3 inches long and 7 or 8 years of age. Cold antarctic winters bring more pack ice and reduce the silverfish populations under the ice. Seven or eight years of cold winters result in a scarcity of silverfish of the right age, and skuas fail to breed. Seven years after a warm winter, the skua population soars.

Wetlands and Waterfowl

A vast region of prairie wetlands exists across the upper Midwest and southern Canadian Prairie provinces (Figure 3). This is the single most important breeding area for waterfowl on the North American continent. According to Batt et al. (1989), these marshes produce 50% to 80% of the continent's total duck population. They are also home to a large variety of nongame birds and quite a few mammal species. The annual waterfowl production is greatest when the number of ponds and wetlands is greatest. It is possible that greenhouse warming may decrease both water depth and the number of ponds holding water in spring and summer, according to Poiani and Johnson (1991). They write: "The number of seasonal and temporary ponds in particular would be expected to decrease. Waterfowl may respond by migrating to different geographical areas, relying more heavily on semipermanent wetlands but not breeding, or failing to renest as they currently do during periods of drought."

Prairie wetlands are relatively shallow and include temporary ponds, seasonal ponds that hold water from spring until early summer, semipermanent ponds that hold water throughout most growing seasons, and large permanent lakes. Annual water levels fluctuate considerably in response to climate variability. Climate controls the water levels and the quality of the vegetation. The quality of the habitat for breeding waterfowl depends on the mix of prairie wetlands types. The birds need a balance of open water and wetlands covered by emergent vegetation. Temporary and seasonal ponds provide abundant early food resources and are much used by dabbling ducks in the spring. Semipermanent ponds provide food and nesting areas for birds as seasonal wetlands dry.

Poiani and Johnson used the GISS $2 \times CO_2$ model projections to estimate the expected climate change in the northern Great Plains. They projected that the mean summer temperature will increase by 2.2° C to 4°C and the mean winter temperature by 5.3°C to 6.6°C. For precipitation they used the figures of Manabe and Weatherald (1986), which project a 30% to 50% decrease in summer soil moisture levels in the Great Plains for a $2 \times CO_2$ climate; however, they recognized the greater

FIGURE 3   Prairie wetlands are the most important breeding grounds for waterfowl on the North American continent, as well as being home to many species of mammals. Greenhouse warming could reduce the depth and number of seasonal ponds such as this one. (Photo by Robert Stottlemyer, Biological Photo Service.)

uncertainty associated with the projection of precipitation. Poiani and Johnson write: "It is unclear how changes in temperature and wetland hydrology will influence primary production, decomposition, nutrient and mineral cycling, and food-chain dynamics. To understand fully the impact of global warming on waterfowl habitat, potential changes in these processes must be determined."

Lower water levels during greenhouse warming would favor the establishment of emergent vegetation, and would thus reduce the amount of open water available to waterfowl. Fewer high-water years would allow the emergent plants to remain and not be flooded out. Wetlands that currently have a high ratio of vegetation cover to open water may become completely closed with plants. Wetlands in basins with open water may become more balanced for wildlife. Extended droughts would dessicate mudflats, allowing the germination of annual plant species and delaying the establishment of emergent cover. Poiani and Johnson modeled the effects of climate change on the vegetation and character of these waterfowl habitats. Their model properly reproduced the present cover-to-water ratio when using the climate of the past decade. Their model simulation of changes projected under global warming confirmed their earlier expectation that emergent cover in semipermanent wetlands will increase considerably and thereby lower waterfowl productivity. In fact, they found a significant decline in waterfowl habitat quality, from a nearly balanced cover-to-water ratio to a completely closed basin with no open-water areas. They concluded with a recommendation for much improved modeling in the future.

Beyond these wetlands of the Great Plains, there is concern for *all* wetlands along the waterfowl migration routes, from California to Kansas to the Atlantic seaboard. These wetlands serve as the birds' feeding stations along their flyways (Figure 4). Delaware Bay is a critical stopover along the eastern flyway. Cheyenne Bottoms, Kansas, has been their major layover in the central flyway. Along the western migration route, Kesterson National Wildlife Refuge and Mono Lake in California, and Stillwater National Wildlife Refuge east of Reno, Nevada, are essential stopping places. These wetlands are already shrinking in size and some are growing saltier—a condition likely to be exacerbated by greenhouse warming.

Mammals

Both birds and mammals are "warm-blooded" animals, or **homeo-therms**. Warm-blooded mammals are capable of adapting to considerable climate variation. They are mobile enough to adjust their range boundaries at a rate entirely in synchrony with the rate of climate change, provided, of course, that their food supply is available. There is a great amount of evidence for their having done so during past climate change events. The pika, a small mammal of the alpine meadows of the western United States mountains, was found at much lower elevations during the last Ice Age. Pikas feed entirely on grasses and do not occupy the boreal forests. As the climate warms during the coming century, pikas will move higher up mountains, limited only by the summit; but on some of the shorter peaks even the summit will be too warm. Eventually, as the boreal forest advances upward on the taller peaks, the pika habitat will be reduced.

Graham (1992) points out that eastern chipmunks, prairie dogs, and boreal bog lemmings all lived together in southwestern Iowa 24,000 years ago. As the climate warmed during the Holocene, only the lemmings migrated northward. Prairie dogs and eastern chipmunks contracted their ranges. Graham writes: "Range, population, and community are constantly changing. Animals migrate at different times, different rates, and in different directions, all for different reasons. I would expect global warming to produce a reassortment of communities

(A)

(B)

FIGURE 4 If wetlands along bird migration routes were to disappear, it is extremely unlikely the bird populations could survive. (A) Snow geese forage in the coastal wetlands of Chincoteague National Wildlife Refuge in Virginia. (B) On the west coast of North America, Mono Lake, California is a crucial link along the flyway. (A, photo by Lars Egede-Nissen; B, photo by Lara Hartley. Courtesy Biological Photo Service.)

and habitat types. Some species might go extinct, but it is difficult to predict which ones."

Climate Warming and Mammal Distributions

When one looks at the distribution maps for North American mammals, one finds a northern boundary in the Great Lakes region for the following species: opposum, least shrew, longtail weasel, badger, gray fox, thirteen-lined ground squirrel, eastern gray squirrel, eastern fox squirrel, southern flying squirrel, white-footed mouse, prairie vole, pine vole, and eastern cottontail rabbit. Most of these mammals would extend their ranges into southern Canada north of the Great Lakes with a warming of the mean annual temperature by 3°C, which is an isotherm displacement of about 250 kilometers. To obtain a more precise determination of these animal displacements, one needs to consider the food requirements for each species and whether or not the particular vegetation, insects, or other food needed would be available farther north.

The mammal species whose southern range boundary is in the Great Lakes region include the arctic shrew, northern water shrew, pigmy shrew, marten, fisher, shorttail weasel, river otter, northern flying squirrel, beaver, heather vole, boreal redback vole, yellownose vole, woodland jumping mouse, porcupine, and snowshoe hare. All of these would also be displaced northward by as much as 250 kilometers during the coming century with the greenhouse warming. Many could disappear entirely from the Great Lakes region. Certainly there would be lags in the displacement of some of these species; once again their food sources would be a determinant. The distributions of some other mammals with southern range limits in the Great Lakes region are largely limited by human activities. These include elk, moose, black bear, gray wolf, lynx, and bobcat.

Scientists are monitoring populations of animals over time to see if they can detect shifts in their ranges as possible responses to greenhouse warming. One of the best places for doing this is in northern Michigan, which is near the ecotone between the boreal and northern hardwood forests. Philip Myers (private communication) studied records of trappings of the boreal redback vole at the University of Michigan Biological Station, Cheboygan County, Michigan. This small rodent was plentiful in the vicinity of the Biological Station prior to the 1940s, but in six years of trapping since 1985 he has caught only one specimen. This appears to be an example of a range shift caused by climate change, since other than the change in temperature, there appears to have been no habitat deterioration. Likewise for the woodland deer mouse, which was plentiful around the Biological Station earlier in the century but is very rare there now. The opposum has migrated northward during the last half century and is now found near Munising in the upper peninsula of Michigan.

Reptiles

Reptiles are "cold-blooded," or **poikilothermic**; they are also **ecto-thermic** since their body heat is derived mostly from the thermal and solar environment; this is in contrast to the endothermic birds and mammals, in which body heat is obtained primarily as metabolic energy from food.

When I first began to work on the energy budgets of animals, I had an idea that I could predict the climate conditions in which an animal must live if I knew its physical and physiological characteristics—i.e., body size, shape, color, insulation quality, metabolic rate, evaporative water loss rate, body temperature, and the variation of these properties with environmental conditions. The first analysis of this challenging problem appeared in Porter and Gates (1969). Our original analysis was applied to about a dozen different animals, including homeotherms and poikilotherms. This work, along with many other publications by my students and me, formed the basis of a new subdiscipline of ecology called "biophysical ecology" or sometimes "mechanistic ecology." Its approach to ecology is quantitative, analytical, mathematical, and based on sound physical, chemical, and physiological principles. The methodology is described in detail by Gates (1980). The potential of this technique for projecting the responses of animals to climate change is enormous.

One group of reptiles, the lizards, generally control their body temperatures by behavioral means that regulate the amount of radiation absorbed—e.g., basking on a sun-drenched rock to raise their body temperature first thing in the morning, or moving into the shade when the midday temperature becomes too high—and through convection and conduction (but seldom through the evaporation of water). Dunham (1993) has used the techniques of biophysical ecology to predict how a change in climate might affect a population of lizards, specifically *Sceloporus merriami*, a small lizard that lives in the Chihuahan Desert of northern Mexico and west Texas. A great deal is known about this lizard's demography, physiology, and behavior as well as the microclimate it inhabits.

Individuals of *S. merriami* have a range of acceptable body temperatures, from 26.9°C to 37.8°C, with an average equilibrium body temperature of 32.2°C. Dunham projected increases of both 2°C and 5°C in the ambient air temperatures and concluded such climate warmings would raise the equilibrium body temperature by about 1.6°C and 4.1°C, respectively. Even under present-day conditions, extreme heat dramatically restricts the amount of time the lizards can spend in foraging, courtship, and other activities during the summer months. Climate warming will further restrict activity times. Even if lizards can remain active at the highest end of their body temperature range (37.8°C), activity times will be severely limited. Lizards maintaining this high body temperature will incur additional maintenance costs because metabolic rates increase with temperature, as does water loss. The demand for more water requires the animal to eat more food—but now there is less time for that. With less time available for feeding, reproduction could be reduced or even halted. Dunham points out that even if survival rates were to increase (due to warmer winter temperatures), reduced reproduction will almost certainly drive the population to extinction, even under the most favorable of scenarios. He then goes on to discuss the limitations of this methodology in making accurate predictions of a population's response to climate change.

Amphibians

Amphibians are cold-blooded; they regulate their body temperatures by their position in the environment and by body moisture loss. Some frogs cool their body temperature below the ambient air temperature by evaporation, or warm it above the ambient temperature by basking in the sun. Amphibians must to live in or around water or in extremely humid environments. A climate change that results in warmer, and particularly drier, conditions would spell trouble for many amphibians. Spotila (1972) described the ecology of terrestrial salamanders and showed how they can adapt to warmer temperatures when properly acclimated and when adequate moisture is available. Salamanders acclimatized to 15°C preferred relative humidities of from 50% to 90%. Many terrestrial salamander species are active at night when relative humidity is high. On dry nights they remain in their burrows with only their heads sticking out. Salamanders have the ability to absorb moisture directly from soil.

Pounds (1991) studied populations of the golden toad in Costa Rica's Monteverde Cloud Forest Reserve. As late as 1987 there were thousands of golden toads mating and laying eggs along streams in the reserve. Four years later this rare amphibian had totally vanished from the region. Precipitation during May 1987 was 64% below average, the lowest ever recorded for that month. Usually early rains in May and June help amphibians recover from the dry season moisture stress in November and December, but the 1987 dry season was exceptionally dry. Except when they emerge to mate, golden toads live in underground burrows that are close to the water table. During the drought the water table plummeted, springs slowed to a trickle, and aquifer-fed streams dwindled. Groundwater normally feeds the streams sufficiently to keep them running during the dry season and to maintain moisture along their mossy banks. Here is also found the harlequin frog, but these disappeared along with the golden toad. Wet streamside patches are crucial to the survival of the harlequin frog, which during the dry season retreats into crevices in the ground, or gathers in moist pockets. Just before these amphibian populations crashed, harlequin frog females emerged in record numbers in search of water. Pounds considered the odds quite low that both species would disappear at the same time purely by chance. Since the amphibians disappeared very suddenly, the drought may have killed the adults and their tadpoles or eggs.

Fish

Fish are cold-blooded and control their body temperatures largely by movement to cooler or warmer water. They swim higher or lower in a lake, or upstream or downstream in a river, seeking a suitable temperature. Temperature controls feeding, metabolism, digestion, sex life, growth, and behavior. Warm-water fish cannot digest food and grow at low temperatures. Deep-lake fish cannot tolerate wide ranges of temperature. Fish adapted to cold water do not survive warm temperatures; yet fish may acclimatize to temperatures that are somewhat warmer or colder than those of their normal habitat. Studies reported by Andrewartha (1961) show that fish can acclimatize seasonally. The bullhead, *Ameiurus*, acclimatized to a maximum lethal temperature of 29°C during the winter and to 35°C during the summer. Many fish have a very narrow preferred temperature range—often as little as 3°C. Figure 5 shows the preferred temperatures for many fish species, along with the upper lethal limits and the upper limits for growth and for satisfactory spawning. For most adult fish, the preferred temperature is about 7°C below the lethal temperature; the temperature limit for satisfactory reproduction is much lower than the upper lethal limit.

The optimal habitat temperature for a specific species of fish must not only be suitable to the fish but also to the organisms upon which it feeds. The mean annual temperature is not as important to fish as the mean maximum and mean minimum temperatures; the *evenness* of the habitat temperature is especially important. The timing of seasonal temperature with spawning, hatching, and growth is critical to fish success. Newly hatched fish must grow, find food, and make it through the first winter.

Cold-water and warm-water fish at the limits of their ranges are

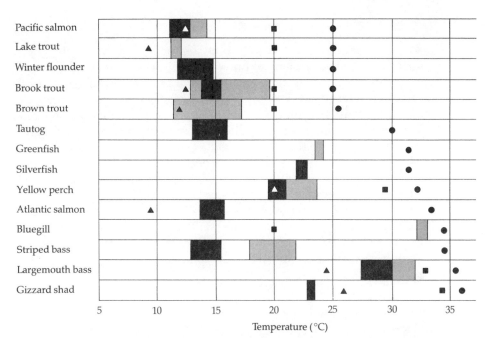

FIGURE 5 **Preferred temperature ranges for some fish species. The light gray boxes are based on measurements made in the field; the black boxes are the preferred temperature ranges of younger fish as measured in the laboratory. The black dots are upper lethal temperature limits; squares mark the upper limits for satisfactory growth, and triangles indicate the upper temperature limit for satisfactory spawning. (From Gates, 1985a, after Clark, 1969.)**

excellent indicators of climate change. The arctic grayling died out in Michigan as waters warmed, and the finscale dace has retracted its range by about 160 kilometers to the northwest in Michigan. The orange spotted sunfish, previously found only as far north as northern Ohio, is now found in southeastern Michigan.

The fish of the Great Lakes may be categorized according to their preferred temperatures. Lake trout and salmon are cold-water fish, walleye and yellow perch prefer cool water, and bluegills and bass are warm-water fish. Greenhouse warming could lengthen the growing season for Great Lakes fish and result in an increase in annual fish productivity. If summers become too warm, however, cold-water fish could suffer. Walleye yields in Lake Michigan could increase by 29% to 33% and lake trout could decline by 2% to 6%, according to some experts. Warmer water temperatures in the Great Lakes will increase the productivity of some algae and zooplankton, the basic food source for many fish.

Insects

Insects are cold-blooded and their body temperatures are largely determined by their environment, although many insects can raise their body temperature through muscular activity. The geographic ranges of most insects are determined by temperature. Usually low temperatures are more significant than high temperatures as determinants of their distributions. Some insects are sensitive to humidity and are most active under very dry conditions (or, for other species, in very humid situations). The wind can transport insects over vast distances. Weather and climate may cause large fluctuations of insect populations. The impact of climate change on agricultural pests is discussed in Chapter 7.

There are a number of excellent reviews of insects and climate, including Uvarov (1931), Messenger (1959), Porter et al. (1991), and Stinner et al. (1989). Temperature increases resulting from greenhouse warming may influence insects in the following ways:

1. Extended geographic ranges
2. Increased overwintering success
3. Increased population growth rates
4. Increased number of annual generations
5. Extended length of the development season
6. Changed synchrony in pest–host relations (see Chapter 7)
7. Changed interactions between insects and their pathogens, parasites, and predators
8. Increased insect migration

Generally, warmer temperatures will increase the geographic ranges of most insect species; however, ranges could be diminished if conditions exceed the upper lethal limit or become too dry for a given species. Greenhouse warming is expected to increase winter temperatures more than it does summer temperatures, so the ability of many insects to overwinter will improve; however, there will certainly be exceptions to this. Insects in temperate and arctic regions undergo cold hardening as winter approaches. More frequent warm spells in the midst of winter may cause a loss of cold resistance and increased mortality when frost returns. Many insects overwinter in ponds, muds, or soils. A change in the depth of frost penetration or of the insulating protection of snow cover may change the survival rate.

The European Pine-Shoot Moth

An excellent example of the effects of temperature on the development and survival of an insect pest is the European pine-shoot moth, which causes heavy damage to stands of red, Scots, and ponderosa pines. Gates (1980) describes the research of Green (1968), one of the most detailed studies of the effect of climate on the life history of an insect. Adults emerge in early summer and lay their eggs on pine shoots. The eggs require two weeks to hatch. The first- and third-instar larvae feed on pine needle tissue. The third-instar larvae feed on buds in August and by early September have bored in for the winter. These larvae, after becoming cold hardened and surviving the winter, molt, begin feeding, and emerge in the spring. The larvae pupate in late May or early June within hollowed-out shoots or buds, and between mid-June and mid-July the adults emerge to begin the life cycle again. All stages of development are temperature dependent. The shoot moth does not occur in any numbers beyond the $-30°C$ mean minimum winter isotherm, except on small trees covered by snow. The insulating effect of snow is excellent, and temperatures of branches under the snow are much higher than that of the air above: the temperature difference between a pine shoot and the air can be as much as $42°C$, depending on the depth of the snow cover. If reduced snowfall or warmer winter temperatures reduces the depth of snow cover, then more mortality could result at times of intense cold outbreaks. On the other hand, warmer winters will move the $-30°C$ mean minimum winter isotherm in North America northward and thus increase the range of the European pine-shoot moth.

Insect Population Growth Rates

Every stage of an insect's development is temperature dependent. The fertility of a female insect may change with temperature. The development rate of insect eggs is often strongly temperature dependent. The growth and development of all instar stages are functions of the ambient temperature. Warmer conditions will usually shorten the life cycle and increase the number of generations during a summer season. Insects that currently produce only one generation per summer may have two or even three generations per summer if conditions warm, particularly if warmth returns earlier in the spring. This effect will be particularly significant at higher latitudes, where the greatest temperature increases are expected to occur. The spruce budworm, for example, may increase with climate warming because its populations are favored by a warm, dry, early spring (Graham et al., 1990). There will be indirect effects of increased temperatures on insects. Many host plants (crops, forests, grasses, etc.) may be stressed by higher temperatures and therefore more vulnerable to insect outbreaks (Rhoades, 1985).

Changes in Interspecific Interactions

Temperature is likely to act differentially on various organisms, and to act at different times of their life cycles. Climate changes could have a considerable impact on insect–predator or insect–pathogen interactions.

Natural controls may become disorganized, allowing an insect species to increase in numbers. According to Porter et al. (1991), "Climatic changes could also create new interspecific interactions and alter existing ones through the expansion of geographical ranges and phenological shifts in species occurrences and abundances."

Migrations of Plague Locusts

The mass migration of locusts is triggered by climate events affecting predation on locusts. According to Walker (1991):

[when there is] a constant high rainfall the predators of locusts maintain development with increasing locust numbers and prevent plagues. Drought-breaking rains, however, lead to plagues, because during the drought the locust predators are greatly reduced and with good rains following drought, there is a considerable lag effect in the relative rates of increase of locusts (from eggs) and their predators.

Walker refers to Uvarov (1931) when stating that although egg development in locusts is cued by temperature, it is not the temperature that determines locust plagues. Here he is referring to the plague locust species of North Africa. The adult plague locusts have the wonderful ability to orient in direct sunshine in order to maintain a preferred body temperature (see Andrewartha and Birch, 1954, and Gates, 1980).

The Great Lakes Ecosystems

The five Great Lakes and their watersheds are the largest source of fresh water in the world. The lakes vary considerably, from the clear, cold, deep basin of oligotrophic Lake Superior to the warm, shallow basin of eutrophic Lake Erie. Over the last few decades the water levels of the Great Lakes have been at both record highs and lows, with a total fluctuation within about one meter. What will the expected climate warming do to the water levels, and how will it affect the important commercial and recreational fisheries? There is evidence that the end of the ice season has come to the Great Lakes increasingly early (10–15 days earlier) over the past 35 years (Hanson et al., 1992).

Climate Change and Lake Levels

The U.S. Environmental Protection Agency sponsored a study of the response of the Great Lakes to climate change. The results of this study are found in Smith and Tirpak (1989) and in Smith (1991). The researchers used climate change scenarios created by GCMs as described in Chapters 5 and 7 of this volume: GISS $2 \times CO_2$, GFDL $2 \times CO_2$, and OSU $2 \times CO_2$. Each scenario started with the average monthly climate for the period 1950 to 1980, used the climate variability for that period, and projected the changes expected under the equivalent of doubled CO_2 concentration. The GCMs all showed increases in mean annual temperatures, GISS by 4°C, GFDL by 6.4°C, and OSU by 3.2°C. Winter temperatures were projected to increase more than summer temperatures, except in the GFDL model, which projected a summer increase of over 7°C. (See Chapter 7 for more discussion of this feature.) Mean annual precipitation increased under all the models; however, summer precipitation decreased strongly in the GFDL scenario, increased in GISS, and showed no change in OSU.

Under the $2\times CO_2$ climate change scenarios, the average levels of all lakes were projected to drop below the historic lows of Lake Michigan and Lake Erie. This decline in lake levels results from higher air temperatures, which reduce the snowpack and increase evaporation. Evaporation does not always increase due to higher temperatures because it is also sensitive to changes in wind speed and humidity. If the average wind speed were to drop and the relative humidity increase sufficiently, then evaporation would drop and lake levels might rise. However, with warmer temperatures it is likely to be windier, and with lower humidity.

Higher air temperatures are likely to mean increased demand for irrigation water and cooling water for power plants. During the record summer heat of 1988, there were loud demands for water diversions from the Great Lakes—diversions down the Illinois and Mississippi Rivers, for example, where power plants were short of water and barges were grounded.

The warmer $2\times CO_2$ scenarios projected reduced ice cover on the Great Lakes. Lake Erie would become almost ice-free in winter, and Lake Superior, instead of having four months of ice cover, would have from 1 to 2½ months of ice per year. Without dredging, ships in Lake Erie ports would need to reduce the weight of their cargoes by 5% to 27% in order to navigate the shallower channels. This would be readily compensated for by a longer ice-free shipping season under each scenario except that of the GFDL model.

Lake Stratification and Fish

During the summer months, as the sun warms the water, the Great Lakes stratify vertically into a warm surface layer and a cool lower layer, separated by a zone of abrupt temperature change called a **thermocline**. This stratification can block the mixing of the well-oxygenated upper layer and the lower, deeper, oxygen-poor layers, while at the same time not allowing nutrients from the lower layers to move upward. Mc-Cormick et al. (1990) found that under global warming Lake Michigan's water would be several degrees warmer at 150 meters deep and would stratify about two months earlier than it does now. This could significantly reduce mixing and have adverse effects on the ecology of the lake. Blumberg and DiToro (1990) analyzed the thermal structure of the central basin of Lake Erie expected under global warming. Higher temperatures were shown to increase the growth of algae, and with increased stratification would lead to lower dissolved oxygen levels.

Magnuson et al. (1990) used the results of McCormick et al. and Blumberg and DiToro to estimate the impact of global warming on Great Lakes fish in southern Lake Michigan and Lake Erie. The study examined (1) changes in the size of optimum habitats for fish, (2) changes in growth and prey consumption, and (3) changes in reproductive success and population size. Laboratory data were used to determine the thermal requirements for each fish species. Not included in the study were species interactions, reduced ice cover, and the effects of lower lake levels on pollutant concentrations. The three classes of fish likely to be most influenced by a significant warming of their habitats are those that do best in cold water (salmon, trout, and whitefish), in cool water (yellow perch and walleye), and in warm water (largemouth and smallmouth bass). Magnuson et al. found that fish would generally benefit from the warmer temperatures. Zooplankton and phytoplankton, the primary food of fish, would increase severalfold, and the thermal habitat for fish during the winter months would be enlarged. The lake trout habitat in southern Lake Michigan might more than double with greenhouse warming. However, whether or not lake trout grew bigger would depend on prey being available. Warmer water temperatures increase the metabolic rates of fish so that they require more food to maintain the same body size. If sufficient prey were available, body size might increase with the warming. Warmer summer temperatures could reduce summer habitat, particularly in areas of shallow Lake Erie or in streams. Spawning areas might be especially vulnerable to warming. Reduced oxygen levels in Lake Erie could cause considerable stress on fish. Warmer waters would make the Great Lakes more attractive to introduced exotic species. These would bring increased competition for food and could cause a loss of current stocks of fish. The lowering of lake levels will also adversely affect marshes and wetlands that serve as breeding and nursery areas for fish and wildlife. Hence, although fish stocks might benefit from greenhouse warming, there are so many unknowns that it is difficult to draw firm conclusions about whether fish production would increase or decrease.

Lakes of the Central Boreal Forest

Some of the GCMs predict that for North America, the greatest effects of greenhouse warming will occur around 48° to 52° north latitude and around 90° to 100° west longitude, where summer temperatures might increase as much as 9°C and soil moisture decrease by up to 50%. The boreal ecosystems of northwestern Ontario may be particularly affected because this region is fairly warm and dry during the summer months and has thin, sandy soils with little water-holding capacity.

Schindler et al. (1990), beginning in 1969–1971, have collected continuous records of weather, hydrology, the chemistry of lakes and their inflow and outflow streams, and the biology of lakes within the Experimental Lakes Area (ELA) in northwestern Ontario. They report on their studies of Lake 239, which is typical of many small boreal lakes on the Precambrian Shield. Their 20 years of records show that air and lake temperatures have increased by 2°C and the length of the ice-free season has increased by 3 weeks. Evaporation rates have been higher than normal and precipitation has been below normal during this period, resulting in decreased rates of water renewal in the lakes. The concentrations of most chemicals have increased in many of the lakes and streams as a result of the diminished renewal rate and of forest fires in the watershed. Populations and diversity of phytoplankton increased in Lake 239, but primary production has shown no consistent trend. The increased water temperatures and deepened thermoclines have caused summer habitats for cold-water organisms with narrow temperature tolerance, such as lake trout and opposum shrimp, to decline.

These scientists feel that these studies give some preview of the effects of increased greenhouse warming on boreal lakes. When the full warming effects of the $2 \times CO_2$ climate change are realized by the mid- to late twenty-first century, the habitats of many cold-water species, including some of the world's most valuable fisheries, may be lost.

Sea Level Rise and Wetland Loss

Marshes and swamps occupy an almost unbroken line along the Gulf and Atlantic coasts of North America (Figure 6). These coastal wetlands support a vast array of aquatic and terrestrial organisms. Titus (1988) described these rich low-lying areas as follows:

Many birds, alligators, and turtles spend their entire lifetimes commuting between wetlands and adjacent bodies of water, while land animals that normally occupy dry land visit the wetlands to feed. Herons, eagles, sandpipers, ducks, and geese winter in marshes or rest there while migrating. The larvae of shrimp, crab, and other marine animals find shelter in the marsh from larger animals. Bluefish, flounder, oysters, and clams spend all or part of their lives feeding on other species supported by the marsh. Some species of birds or fish may have evolved with a need to find a coastal marsh or swamp anywhere along the coast. Wetlands also act as cleansing mechanisms for ground and surface waters.

FIGURE 6   Salt marsh and tidal channels mark the coastal wetlands of South Carolina (shown here) as well as most of the rest of the Atlantic and Gulf Coasts of North America. These coastal ecosystems are vital to human society and to countless species intertwined in their complex food webs. Water purification is only one of the functions such wetlands perform, and their development by oil and real estate interests must be carefully monitored if their quality is to be maintained. The threatened rise in sea level under global warming scenarios is a significant danger. (Photo by Cub Kahn, Terraphotographics.)

The coastal wetlands are of vital importance to human society in every conceivable manner. Gates (1985a), considering the impact of near-shore, or onshore, oil development along the Gulf Coast, described the ecology as follows:

Food webs here are based on phytoplankton. Many burrowing organisms (filter-feeder types) are found in the bottom muds. The region is an important spawning ground for sport and commercial fishes (menhaden, Atlantic croaker, and mullet) and for invertebrates (brown and white shrimp). This zone is also the year-round habitat for ocean sunfish, oarfish, swordfish, king mackerel, seals, and whales. The beach spawning area is an intertidal region continually buffeted by waves, tides, and currents. It possesses rich floral and faunal communities. Many mollusks, annelids, and crustaceans are found there, as well as juvenile pompano, Gulf kingfish, banded drum, longnose killifish, and rough silversides. It is used as a nesting area by boobies, brown pelicans, terns, gulls, and other shore and water birds.

These two descriptions give an idea of the diversity of life harbored in the coastal zones. Actually these zones are much more complex than described here, since they run from saline intertidal marshes to brackish intertidal wetlands to freshwater wetlands farther inland.

Global warming will produce a rise of sea level by expanding the volume of seawater, melting mountain glaciers, and causing ice sheets in Greenland and Antarctica to melt or slide into the oceans. Titus (1988) reports estimates of sea level rise by 2025 to be 10 to 21 cm, by 2075 to be 36 to 191 cm, and by 2100 to be 57 to 368 cm. Because much of the east coast of the United States is sinking, the relative rise at a particular location may be greater. Sea level rise will affect coastal wetlands by inundation, erosion, and saltwater intrusion. Some wetlands will be converted to open water; in others, the type of vegetation will change, although they will remain wetlands. If the rise of sea level is sufficiently slow, wetland vegetation can trap sediments or build peat and prevent the rising salt water from disrupting the wetland.

Periodic flooding is a characteristic of coastal marshes, but increased flooding will alter these ecosystems. Salt marshes extend seaward to approximately the elevation flooded at mean tide, and inland to the area flooded by spring tide (defined as the highest astronomical tide every 15 days). Salt marsh plants, such as *Spartina*, are notable for their varying tolerance of salinity. Coastal wetlands flooded once or twice daily support "low marsh" vegetation; areas flooded less frequently support "high marsh" plants. Titus writes:

The natural impact of a rising sea is to cause marsh systems to migrate upward and inland. Sea level rise increases the frequency and/or duration of tidal flooding throughout a salt marsh. If no inorganic sediment or peat is added to the marsh, the seaward portions become flooded so much that marsh grass drowns and marsh soil erodes: portions of the high marsh become low marsh; and upland areas immediately above the former spring tide level are flooded at spring tide, becoming high marsh. If nearby rivers or floods supply additional sediment, sea level rise slows the rate at which the marsh advances seaward.

In many coastal areas the land above the marsh is steeper than the marsh, so a rise in sea level inundates the marsh and causes a net loss of acreage. In Louisiana, vast marsh areas are within one meter of sea level, with very narrow ridges in between the marsh and the sea and very little low adjacent upland. A one-meter rise of sea level would drown most of the Louisiana wetlands without creating a great deal of new marshes. In New England and along the Pacific coast, high cliffs prevent the movement of seawater inland, so the marshes would drown and not be replaced. **Tidal range**—the difference in elevation between the mean high tide and mean low tide—determines the vulnerability of a coastal area to sea level rise. Coastal wetlands are generally less than one tidal range above mean sea level, although there is a great variation in tidal ranges throughout the United States: over four meters in Maine, less than two meters along the mid-Atlantic seaboard, and less than one meter along the Gulf of Mexico.

Generally, coastal marshes will not keep pace with the expected rate of sea level rise, but three possible exceptions are the marshes found in river deltas, tidal inlets, and on the bay sides of barrier islands. It is possible that sediment deposits in river and tidal deltas will allow marshes to keep up with rapid sea level rise. Barrier islands tend to migrate landward as storms wash sand from the ocean-side beach to the land-side marsh. However, it is not clear whether this mechanism will prevail or whether coastal dunes will simply disintegrate with rising sea level. Seawater intrusion could produce large vegetation changes on inland marshes. Freshwater marshes are often tens of kilometers or more inland and their elevations are nearly the same as those of the saline wetlands. Rising sea level, often combined with droughts, allows seawater to penetrate inland. Many of the extensive cypress swamps in Louisiana, Florida, and South Carolina may convert to open water, a process already occurring in Louisiana.

Rough estimates have been made of the potential loss of United States coastal wetlands (Titus, 1988). There are an estimated 2.79 million hectares of coastal wetlands at present. Under the lower sea level rise scenario (a rise of less than 1 meter), 1.3 million hectares might be lost by the year 2100. It is possible that 0.5 million hectares could be created by sedimentation if human activities, such as coastal buildings, do not interfere. Under the higher sea level rise scenario (a rise of more than 2 meters), 2.3 million hectares (81%) might be lost, and 0.8 million hectares of marsh created if human barriers do not get in the way. If rates of sediment deposition do not create new marshes and greenhouse warming continues, the United States will likely lose most of its coastal wetlands by the year 2100.

Ozone Depletion, Ultraviolet Radiation, and Life

Sunlight is essential for life, yet in excess it can be harmful to life. It is the ultraviolet portion of sunlight that is most detrimental to plant and animal tissues, but even some ultraviolet radiation is beneficial. For example, vitamin D is produced in the skin by the action of ultraviolet rays. If the human diet supplies too little vitamin D, the result is rickets and various bone deformities. An adequate amount of sunlight generates sufficient vitamin D to make up for any dietary deficiency. But too much ultraviolet radiation can produce cell mutations, DNA damage, skin cancer, premature aging, and even damage to the human immune system. A delicate balance between the sun's beneficial wavelengths and its destructive rays exists for life at the Earth's surface. The ozone in the upper atmosphere filters out the most destructive of the solar wavelengths and therefore is critical for maintaining this balance. Any ozone-damaging process occurring in the stratosphere is of immediate and critical concern to human well-being.

The world is sometimes a crazy place. As smart as we think we are today, we are still caught short by environmental surprises. The discovery in 1985 of a stratospheric ozone hole above Antarctica was one such surprise—and this came after years of concern about increasing *tropospheric* ozone concentrations within and near large urban areas. The lower-atmosphere ozone increase results from hydrocarbon emissions that generate photochemical smog. The upper-atmosphere ozone decrease relates to the release of chlorofluorocarbons and other compounds into the stratosphere. As early as 1971, Johnston (1971) had called attention to the possible destruction of stratospheric ozone by nitrogen oxide catalysts emitted in the exhaust of supersonic airplanes. But only during the past decade has the stratosphere received the sophisticated scrutiny necessary for us to understand its composition and potential for change. As many as 80 chemical reactions must be taken into account when modeling the stratosphere. (See Lal and Holt, 1991; McElroy and Salawitch, 1989.)

Ozone Formation and Destruction

Ozone (O_3) forms naturally in the stratosphere when sunlight splits the oxygen molecule (O_2) into two single oxygen atoms. These oxygen atoms can then collide with oxygen molecules in the presence of a third substance to form ozone. These reactions are written as follows:

$$h\nu + O_2 \rightarrow O + O \tag{1}$$

$$O + O_2 + M \rightarrow O_3 + M \tag{2}$$

where $h\nu$ represents sunlight and M represents H, OH, H_2O, HO_2, or another substance.

Ozone also can be destroyed by sunlight and collision with oxygen atoms. These reactions can be written:

$$h\nu + O_3 \rightarrow O + O_2 \tag{3}$$

$$O + O_3 \rightarrow O_2 + O_2 \tag{4}$$

The distribution of ozone in the stratosphere is maintained by a balance between its production and its loss, as well as by its movement from regions of production to regions of net loss. This movement results from variable winds, which cause daily, seasonal, and interannual var-

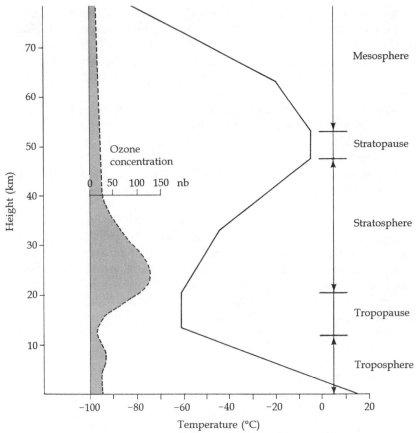

FIGURE 7   **Vertical profiles of ozone concentration and atmospheric temperature.**

iations in ozone concentrations. The result is that ozone is distributed in the stratosphere with a maximum concentration at an altitude between 20 and 25 km, as shown in Figure 7.

Ozone absorbs ultraviolet radiation very effectively, with the result that the uppermost levels of the stratosphere are heated by the sunlight. But above these altitudes, where little ozone remains, the atmosphere becomes colder again. Ozone and its interaction with sunlight is primarily responsible for the presence of the stratosphere as a distinct region of the atmosphere. Ozone is an effective greenhouse gas and it, along with carbon dioxide, radiates infrared energy to outer space, thereby cooling the stratosphere. The fact that the stratosphere has a warm top and a very cold base gives it dynamic stability. Air within it will not rise from bottom to top of its own buoyancy, as it does in the lower atmosphere. The very low temperature of the stratospheric base, the tropopause, creates a cold trap for water vapor. Therefore very little water vapor enters the stratosphere from below and the air above the tropopause is extremely dry. All of this is important to the rest of our story.

The Role of Nitrogen

High-temperature combustion, lightning, and soil microbial activity produce oxides of nitrogen (NO_x). These gases diffuse into the stratosphere, where they become significant to ozone chemistry. It was while investigating the environmental impact of high-flying supersonic aircraft that Johnston (1971) realized that the nitric oxide (NO) released in the exhausts of these planes could destroy stratospheric ozone. Nitrogen oxides can break up ozone molecules in a catalytic reaction that can be written as follows:

$$NO + O_3 \rightarrow NO_2 + O_2 \tag{5}$$

$$NO_2 + O \rightarrow NO + O_2 \tag{6}$$

Chlorine and Bromine

Now the plot thickens. In 1974 Mario Molina and Sherwood Rowland suggested that chlorofluorocarbons could also destroy stratospheric ozone. CFCs in the stratosphere are broken up by sunlight into their constituent molecules—chlorine (Cl), fluorine, and carbon—leaving the chlorine free to catalyze the breakup of ozone in a manner similar to that of NO:

$$Cl + O_3 \rightarrow ClO + O_2 \tag{7}$$

$$ClO + O \rightarrow Cl + O_2 \tag{8}$$

Compounds similar to the chlorofluorocarbons contain bromine (Br) instead of chlorine and react in the same way.

The Ozone Hole

Farman et al. (1975), members of the British Antarctic Survey, were reviewing atmospheric ozone measurements made at Halley Bay, Antarctica, and discovered a startling decrease in the total amount of ozone. They found that since 1968 the ozone amount had a downward trend. The British team had been making these measurements since 1957 using a Dobson spectrophotometer, which measures the total amount of ozone in a vertical column extending upward from the instrument. The daily column value in October prior to 1968 was greater than 300 DU

(Dobson Units), but by 1984 had fallen to less than 200 DU. In October 1991 the column value was less than 150 DU. Although this is a 50% loss from the pre-1968 amount, the loss within the hardest-hit layer of the lower stratosphere, between 15 and 20 kilometers in altitude, is 95%. This ozone depletion, known as the **ozone hole**, occurs over most of Antarctica during the austral spring for a period of about 6 weeks. It represents the destruction of about 3% of the total world stratospheric ozone—a truly catastrophic loss. Farman and colleagues immediately thought of chlorine as the culprit. But when atmospheric chemists tried to model the chlorine released from CFCs accumulating in the atmosphere, they could not get anything like the rate of ozone destruction taking place over Antarctica. That is because ordinarily the breakup of CFCs would be followed by chemical reactions of chlorine monoxide (ClO) with nitrogen dioxide (NO_2) to form chlorine nitrate ($ClONO_2$), as follows:

$$ClO + NO_2 + M \rightarrow ClONO_2 + M \qquad\qquad (9)$$

This reaction ties up the chlorine in chlorine nitrate, which is a non-reactive chemical species, also known as a "reservoir species." This would leave very little chlorine in reactive forms such as Cl or ClO. Something was very wrong.

Polar Stratospheric Clouds

Antarctica is a most unusual part of the world. Due to intensive radiative cooling in winter, coupled with the rotation of the Earth, a massive, highly stable vortex of cold air develops over Antarctica during the winter months. Within this vortex are extensive cloud formations referred to as polar stratospheric clouds (PSCs). These clouds are unusual because of their extremely low temperatures, less than $-78°C$. They are composed of small, frozen nitric acid trihydrate particles ($HNO_3 \times 3H_2O$). During the cold polar night, reactions occur on the surfaces of PSC particles that repartition chlorine from chlorine nitrate into other reservoirs, as follows:

$$ClONO_2 + HCl(s) \rightarrow Cl_2 + HNO_3(s) \qquad\qquad (10)$$

$$ClONO_2 + H_2O(s) \rightarrow HOCl + HNO_3(s) \qquad\qquad (11)$$

$$N_2O_5 + HCl(s) \rightarrow ClNO_2 + HNO_3(s) \qquad\qquad (12)$$

$$N_2O_5 + H_2O(s) \rightarrow 2HNO_3(s) \qquad\qquad (13)$$

where (s) denotes the solid phase.

These reactions tie up nitrogen in solid nitric acid, $HNO_3(s)$, removing it from the gas phase where it otherwise would be destructive to ozone. Also, denitrification of the stratosphere occurs as sedimentation, the sinking of particles, takes place. During the Antarctic winter the PSCs are responsible for a significant reduction in the NO and NO_2 levels. Ozone levels thus are high throughout the polar night. But in springtime, the sunlight breaks HOCl and Cl_2 into the reaction forms Cl and ClO (and, similarly, the bromine compounds into Br and BrO). The low NO_2 levels allow the amount of free chlorine to remain high. Spring sunlight also evaporates the PSCs, realeasing HNO_3. If all of this HNO_3 were simply released into the atmosphere, it would be broken up by sunlight, raising NO_2 levels, and ozone destruction would be mitigated by the reaction of NO_2 with Cl to form reservoir species. But HNO_3 removal by sedimentation eliminates this sink for reactive chlorine, and as a result ozone destruction increases.

Destruction of ozone by chlorine accounts for about 75% of the ozone loss; that by bromine, only about 25%.

Observational Evidence for the Role of CFCs

All the theoretical chemistry in the world will not prove that CFCs are a prime culprit in the destruction of stratospheric ozone without confirmatory observational evidence in situ. This is exactly what James Anderson and his colleagues found (see Anderson et al. 1991). They outfitted a medium-range, high-altitude (19 km) ER-2 reconnaissance aircraft and a long-range, medium-altitude (11 km) DC-8 aircraft with equipment to measure the atmospheric concentrations of ClO, BrO, and O_3. Ten flights were made from Punta Arenas, Chile into the polar vortex and out again during late August and most of September 1987. Outside the polar vortex, ClO levels were 2 to 5 parts per trillion, every bit as low as measured at mid-latitudes. Then across the vortex edge, ClO levels jumped to two orders of magnitude higher than those observed outside the vortex. Furthermore, companion measurements showed that where ClO levels were high, levels of nitrogen oxides and nitric acid were very low. However, most striking was the observation that where ClO concentrations increased, O_3 concentrations decreased. This in itself does not prove cause and effect, but along with the theoretical chemistry it is pretty convincing. As Anderson wrote, "When taken independently, each element in the case contains a segment of the puzzle that in itself is not conclusive. When taken together, however, they provide convincing evidence that the dramatic reduction in column-integrated O_3 over the Antarctic continent would not have occurred had CFCs not been synthesized and then added to the atmosphere."

Ozone Depletion Worldwide

In early spring the Antarctic stratosphere contains an intense westerly circulation pattern—the polar vortex—within which the very low levels of ozone just described are found. This vortex is surrounded by a ring of high ozone concentration in mid-latitudes. During summertime the Antarctic vortex breaks down, and the ozone-depleted air escapes into the lower latitudes and into the troposphere. Record low ozone levels were found over southern Australia and New Zealand during October 1987 (Atkinson et al., 1989). This appears to have been the result of direct transport of ozone-depleted air from Antarctica.

Now the story moves to the North Polar region. Hofmann and Deshler (1991) reported reductions in stratospheric ozone levels in the Arctic polar vortex above Kiruna, Sweden from 8 January to 8 February 1990 as measured with balloon-borne instruments. Reductions of 25% occurred in the 22-km altitude region of the stratosphere during part of the period of measurment. Reductions in ozone seem to be related to the time air parcels spent at temperatures less than $-80°C$. The winter polar vortex in the Northern Hemisphere is warmer and more dynamically disturbed than that in Antarctica. In fact, in January 1992 there was a sudden warming of the Arctic stratosphere, and fewer PSCs formed than usual. The ozone losses peaked at less than 10%, rather than the 50% typical of the Antarctic. As greenhouse gases increase in the atmosphere, tropospheric warming will be accompanied by stratospheric cooling, and this cooling will increase the probability and persistence of PSCs in the Arctic and will favor a long-term trend of decreasing ozone amounts.

Although dramatic ozone losses have not been found in the Arctic, a gradual long-term downward trend has been reported at northern mid-latitudes (roughly between Seattle and New Orleans). NASA's satellite-borne Total Ozone Mapping Spectrometer (TOMS) indicated ozone losses of 4% to 5% per decade in that region. Measured during the winter months, the rate of loss is as high as 8% per decade. Globally, ozone is decreasing 2.3% per decade. What mechanisms could be driving these mid-latitude losses? Laboratory experiments have suggested another mechanism beside PSCs for catalyzing the destruction of ozone during winter months. Some of the reactions that destroy ozone at the poles may be taking place on tiny droplets of sulfuric acid, like those that form the thin haze being found in the stratosphere. Modeling of these reactions in the mid-latitude stratosphere seems to confirm the possibility that not only could they result in serious wintertime ozone losses in the mid-latitudes, but losses during the summer as well (see the report by Appenzeller, 1991).

The ER-2 reconnaissance aircraft has flown into the sulfuric acid cloud from the Mount Pinatubo volcanic eruption along flight paths from Maine to Cuba. This and other measurements have found unexpected levels of ClO at mid-latitudes and even over the tropics. Simultaneously, stratospheric haze particles are widespread. These particles include natural aerosols of sulfuric acid, including the Pinatubo cloud, and perhaps ice crystals over the tropics. Until recently, only the PSCs had been implicated in ozone destruction. Now it is believed that these other particles are promoting the ozone-destroying chemistry. It is possible that these particles are acting catalytically to immobilize oxides of nitrogen, which otherwise would be converting ClO into reservoir form. Jones (1989) summarizes studies that show the ozone concentration between 10° and 50° north latitide to have been depleted by about 15% following the eruption of El Chichón in 1982. The sulfuric acid cloud from Mount Pinatubo will persist for a few years and then fade, but chlorine concentrations in the atmosphere are still increasing and will not peak until after the turn of the century.

Ultraviolet Radiation Increases

If changes in stratospheric ozone levels only affected the dynamics of the atmosphere, it would be serious enough. However, decreases in ozone concentrations may allow increases in the intensity of ultraviolet radiation reaching the Earth's surface, with serious consequences for organisms, including humans. The spectral band of most concern biologically is UV-B radiation, at wavelengths from 280 to 320 nm, and to a much lesser extent UV-A radiation, from 320 to 400 nm. The shortest and most damaging UV wavelengths, UV-C (200 to 280 nm), are absorbed strongly by the atmosphere so that essentially none reach the Earth's surface.

Biological responses to ultraviolet radiation depend on the product of the intensity of solar radiation at each wavelength and the action spectrum of an organism—its relative sensitivity to damage by sunlight—at each wavelength. The product is referred to as the "biologically effective irradiance." Figure 8 exemplifies the significance of changes in UV-B irradiance. A typical biological action spectrum for damage to DNA, and closely approximating that for the reddening of human skin, is shown in the figure. Also shown is the mid-latitude, solar-noon irradiance in summer for a standard column ozone amount of 0.32 atm cm. The resulting biologically effective irradiance is also shown. If there were to be a 50% reduction in the column ozone amount, the solar irradiance would increase as shown and the biologically effective irradiance would be amplified. (A review of the action spectra for higher plants and the solar effectiveness is given by Coohill, 1989.)

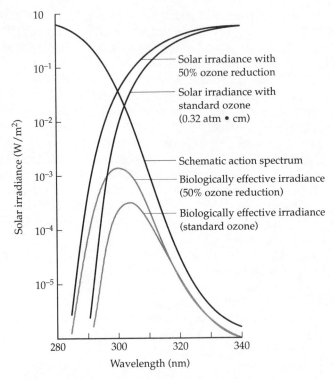

FIGURE 8   **Solar irradiance with standard ozone thickness (0.32 atm cm) and with a 50% ozone reduction. A typical biological action spectrum is shown. The biologically effective irradiance is the product of the solar irradiance and the action spectrum.**

Solar UV-B radiation increases with decreasing latitude and with increasing elevation, as well as from winter to summer and from dawn to midday. Very few systematic routine UV-B measurements have been made at various locations on Earth over an extended period; Urbach (1969) summarizes most of the early measurements. Although these early studies did not detect an increase in UV-B radiation corresponding to the stratospheric ozone depletion from 1969 to 1986 in the Northern Hemisphere, Blumenthaler and Ambach (1990) report a slight increase of 1% per year above the Jungfraujoch High Mountain Station in the Swiss Alps since 1981. This increase corresponds with a measured decrease of column ozone amount over Davos, Switzerland of 4% from 1969 to 1988. Particularly high values of UV-B radiation were measured in the spring of 1983, after eruptions of the El Chichón volcano, and there was a concurrent decrease in total ozone amount. If Blumenthaler and Ambach's results are indicative, then the UV-B flux will not increase as rapidly with ozone reduction as earlier believed. Earlier estimates suggested that for a 1% decrease in the column ozone there would be a 2% increase in UV-B and a 4% increase in the number of skin cancers.

Even though the ozone concentration above Antarctica has been measured for several decades, there have not been concurrent measurements of UV-B irradiation at the surface there. Frederick and Snell (1988) reported calculations of UV-B irradiance at McMurdo Station, Antarctica during depleted ozone and normal ozone conditions. A total column ozone amount of 315 DU was typical during October over Halley Bay, Antarctica, in years prior to 1968. A column ozone of 250 DU existed over McMurdo in early September 1987, and the lowest value found over Antarctica was below 110 DU, on 5 October 1987. Frederick and Snell found that the lowest ozone amounts observed in October suggested biologically effective irradiances comparable to or greater than those for the same location at the summer solstice, and even comparable to those at the summer solstice at the latitude of Miami. Thus, life forms indigenous to Antarctica experience a greatly extended period of summerlike ultraviolet radiation during the appearance of the ozone hole.

Biological Consequences of Increased Ultraviolet Radiation

The biological effects of light are the result of the absorption of specific wavelengths of light by specific molecules in cells. The result may build useful compounds, as in photosynthesis, or be destructive of cells, as with sunburn. DNA—deoxyribonucleic acid, the molecule that carries the genetic information of the cell—can be damaged by ultraviolet radiation. The electronic structure of DNA is such that it absorbs UV radiation but not visible light. Most living cells possess a multitude of systems for repairing damage to their DNA. One repair mechanism involves photoreactivation, an enzymatic process that uses visible light around 400 nm. There is also a "dark repair" system that does not require light. Plants and animals have also evolved means to protect themselves from damaging UV. Many organisms have protective pigments. Normal human skin develops melanin after exposure to sunlight, and this pigment screens the underlying tissue. Hair, feathers, scales, and shells also protect animals from solar UV. Many animals and even aquatic plants can move into protective niches or deeper water to escape damaging radiation. Usually juvenile stages, such as plant buds and animal egg or larval forms, are most vulnerable to UV-B damage.

The Antarctic Ecosystems

There is great concern for both aquatic and terrestrial ecosystems around and within the Antarctic continent due to the annual occurrence of the ozone hole and the concurrent increase in UV-B radiation. The ozone hole first appears in September, reaches its peak in October, and dissipates during November. However, springtime is when most organisms are just emerging from the extended cold and darkness of winter and are beginning the reproductive time of their life cycles (although some species have their breeding season as late as January, well after the ozone hole has disappeared). The ocean surrounding Antarctica is one of the most productive aquatic ecosystems in the world. By contrast, the terrestrial ecosystems there are quite sparse, containing a total of only about 600 species of lower plants, two species of higher plants, and 150 species of invertebrates. Most of the concern about UV-B damage to these organisms is speculative at the present time, since very few direct observations and measurments have been made.

Marine Ecosystems

Marine life in the ocean's surface waters, where most solar radiation penetrates, may be adversely affected by increased levels of UV-B radiation. The productivity of phytoplankton in the Antarctic marginal ice zone (MIZ) is thought to contribute significantly to the overall productivity of the Southern Ocean. In spring, when the ice melts, a highly stable layer of fresh water forms on top of saltier water underneath. Because of this stratification, the algal blooms that are the base of the marine food chain are restricted to the near-surface waters of the MIZ. Since so much of its primary productivity is in the surface waters, this ecosystem is particularly vulnerable to UV-B radiation. The concern is that if Antarctic phytoplankton are affected, then krill, the tiny animals that feed on them, will be affected in turn, followed up the food chain by fish, seabirds, seals, and whales. The entire chain could be damaged. Most scientists do not believe the food chain will be wiped out, but rather that the ozone hole will lead to a change in species composition as more tolerant species replace the most sensitive ones.

Smith et al. (1992) reported the results of a six-week cruise in the MIZ of the Bellingshausen Sea during the austral spring of 1990. These investigators used a submersible UV spectroradiometer to determine the penetration of UV radiation into Antarctic waters, and an incubation system to quantify the impact of UV-B radiation on phytoplankton at various depths. The ship's route was designed to take it from outside the area underneath the ozone hole to well inside the hole. During the six-week period, September to November, they carried out four north–

south (each 280 km) and three east–west (each 72 km) transects, along which they performed intensive vertical profiles of the ocean water to define its physical, optical, chemical, and biological characteristics. In addition they made a long east–west transect (800 km) to broaden the longitudinal extent of the observations. The submersible spectroradiometer detected UV-B radiation at depths in excess of 60 to 70 meters.

According to Smith et al. (1992), "Phytoplankton have a number of photoreactivation and photoprotective strategies to partially compensate for photoinhibitory effects of different wavelengths of light." Their results demonstrated variation among phytoplankton species in their ability to protect themselves against UV-B irradiation. The inhibition of photosynthesis by UV-B radiation in a species of the phytoplankton *Phaeocystis*, for example, was much greater than that in the diatom *Chaetoceros socialis*. Furthermore, Smith et al.'s results suggested that a test bacterium with no means of DNA repair experienced a 3% loss of growing cells at a column ozone of 350 DU, compared with a 25% loss at 200 DU. These results indicated an upper limit for biological damage caused by increased UV-B within the water column.

Smith et al. used *Phaeocystis* to test the influence of variable UV-B radiation on the productivity of the most abundant phytoplankton communities of the MIZ. Inhibition of photosynthesis was evident in samples incubated at depths down to 25 meters. (At this depth 4% of the radiation is photosynthetically active.) UV-B inhibition of photosynthesis increased linearly with increasing UV-B dose. UV-B inhibition of surface phytoplankton may be greatest during morning hours and may be mitigated by photoprotective mechanisms induced by sunlight during the day. In general, shifts of solar spectral irradiances due to ozone concentration changes alter the balance of spectrally dependent phytoplankton processes, including photoinhibition, photoreactivation, photoprotection, and photosynthesis. A minimum 6% to 12% reduction in primary productivity associated with ozone depletion was estimated during the cruise.

Cullen et al. (1992) report on laboratory experiments in which a diatom (*Phaeodactylum* sp.) and a dinoflagellate (*Prorocentrum micans*) were irradiated with UV-A and UV-B radiation. Measured photosynthetic rates were fit to an analytical model to predict the short-term effects of ozone depletion on aquatic photosynthesis. The results indicated that UV-A (320–400 nm) significantly inhibits photosynthesis by these organisms and that the effects of UV-B (280–320 nm) are even more severe. The model shows that the Antarctic ozone hole might reduce near-surface photosynthesis by 12% to 15% (but less so at depth). Since phytoplankton move vertically through the water column at different rates, and therefore experience different exposures to UV radiation, so the kinetics of photoinhibition and recovery must be considered when calculating total water column photosynthesis.

The Cullen et al. analysis differs from that described previously, by Smith et al. (1992), in which it was suggested that UV-A played an important role in photoprotection against UV-B damage. The length of time of exposure to UV radiation was quite different in the two tests, being much longer in the Smith et al. study.

In 1987 and 1988 Karentz and colleagues (see Karentz et al., 1991) conducted on-site evaluations of ultraviolet exposure and photobiological responses of nine phytoplankton species at Anvers Island, Antarctica. They detected UV-B radiation penetration as deep as 20 meters during maximum ozone depletion. The nine diatom species they studied spanned nearly three orders of magnitude in cell volume and varied considerably in their surface area-to-volume ratios. All showed evidence of photoreactivation and DNA repair at ambient Antarctic temperatures, but with much variation among species. These authors point out that "the ozone hole has existed for more than ten years. Obvious catastrophic events have not been noted. Our results indicate that Antarctic phytoplankton have diverse capabilities for sustaining and repairing UV-induced damage to DNA and as a group are not defenseless against this environmental stress."

Terrestrial Ecosystems

Plants and animals on the land in Antarctica live near the limits of their physiological tolerances and are often stressed by extreme temperatures and water shortages. Some terrestrial organisms have developed protective systems against UV-B radiation and many are heavily pigmented. According to Voytek (1990), studies show that the following physiological characteristics of flowering plants are susceptible to UV-B radiation: "photosystems I and II; carboxylating enzymes; stomatal resistance; chlorophyll concentration; soluble leaf proteins; lipid pools; carbohydrate pools; epidermal transmission (i.e., water retention and CO_2 exchange)." Interesting studies reported by Caldwell et al. (1982) show that Arctic ecotypes and species of plants growing where solar UV-B radiation is very low were consistently more sensitive to UV-B radiation than their counterparts from lower-latitude alpine regions, where solar UV-B radiation is high. This would certainly imply that the two higher plant species that live in Antarctica may be vulnerable to UV-B damage.

Very little seems to be known about the effects of UV-B radiation on arthropods (including insects). Many species apparently are little affected, and those that may be are sensitive primarily during egg and larval development.

In sum, it seems that the greatest impact of environmental change on the Antarctic ecosystems will be from global warming rather than from ozone depletion and ultraviolet damage; however, this is not to suggest that damage to the ecosystems from increasing levels of UV-B radiation will not be serious.

CHAPTER SEVEN

Agriculture, Droughts, and El Niño

ALL ANIMALS ARE driven by their food supply, and human beings are no exception. The world's human population now exceeds 5 billion and is projected to exceed 8 billion by about 2025. Food production has grown steadily since 1960, but the distribution of food is far from equitable. During the 1980s severe droughts and political disruptions interfered with food production and distribution in Africa, Bangladesh, Pakistan, and India. Climate extremes impaired United States agricultural production during the Dust Bowl times of the 1930s and again in 1983, when corn yields fell drastically in the southeast. Again in 1988, a hot, dry summer decreased the U.S. corn yield by almost 40%.

The demand for food increases with world population growth and results in more forests felled, more fields cleared and abandoned, more ecosystems impoverished, more soils depleted, more air and water polluted, more energy, water, and fertilizers used, and more greenhouse gases released to the atmosphere. Climate has always been a determinant of agricultural production. This is as true today as it was in the past and it will be so far into the future. It is essential to have a good understanding of the climate processes affecting crop productivity. This knowledge may not only allow for improved crop management today, but will be crucial for crop management in the future as the world becomes warmer, drier, or wetter—or for that matter, even if the world should get colder.

Nearly all crop models are semi-empirical, in that they are derived from a great deal of field data representing crop response to climate and soil, but with considerable physiological information added. I remember well a discussion I had a decade ago with a leading crop breeder, who when asked if crops might be found that could cope with the climate changes expected with greenhouse warming, answered: "Yes! We are constantly testing crop varieties throughout the world under all kinds of climate and soil conditions and while the greenhouse warming is in effect and the carbon dioxide levels are increasing." He suggested that plant breeders will select crop varieties as conditions merit and will indeed be ready to meet the needs of the future. That, no doubt, is considerably true, but it does not obviate the need for good crop modeling in order to anticipate which crops will grow where in the future and just which varieties might be needed.

Crop Modeling

The U.S. Environmental Protection Agency (Smith and Tirpak, 1989) sponsored a detailed study to evaluate the possible effects of global climate warming on United States agriculture. This was part of the same effort as the study reported in Chapter 5 for U.S. and Canadian forests. Approximately the same climate scenarios were used for the crop study as for the forest study. The crop modeling experiments were distinctly different from the forest modeling calculations, however, because crops have a life span of one year or less, whereas trees may grow for a century or longer. In most respects, there is much more physiological information available for crops than for trees, and even the microclimates of crops are more precisely known and modeled.

The Crop Model Characteristics

Production of corn, wheat, and soybeans takes up about two-thirds of the total U.S. agricultural acreage. The economic value of these crops is equal to that of all other crops combined. For this reason, these are the crops on which modeling efforts have concentrated. The main producing regions for these crops in the United States are east of the Rocky Mountains. Wheat is grown primarily in the Great Plains; corn and soybeans are grown in the midwest region known as the Corn Belt, the Great Lakes states, and the southeast, including the Mississippi Delta region.

The crop models used were tested satisfactorily throughout the world for at least five years. The models used to simulate wheat, corn, and soybeans were CERES-Wheat (Ritchie and Otter, 1985), CERES-Maize (Jones and Kiniry, 1986), and SOYGRO (Wilkerson et al., 1983). The models were designed to predict the growth and yield of different wheat, corn, and soybean varieties for all possible climate and soil conditions and under various management strategies. The model outputs included phenological development of growth stages; development of vegetative and reproductive units; growth of leaves and stems, as well as leaf senescence; biomass production and partitioning; and root dynamics. The models also simulated the effects of soil-water deficit and nitrogen deficiency on photosynthesis and pathways of carbohydrate movement in the plant.

The input variables for these crop models were put into three categories: (1) exogenous variables, which were uncontrollable and possibly stochastic; (2) controllable, or management, variables, which were deterministic; and (3) system parameters, which were coefficients in the analytical expressions. The exogenous variables included daily solar radiation, maximum and minimum air temperature, and rainfall. The controllable variables were the beginning day of the simulation, the day of the year for sowing, the plant population density (number of plants per square meter), row spacing, depth of sowing, timing and amount of irrigation, and amount and frequency of nitrogen fertilization. For most of the climate change simulations, the amount of nitrogen fertilizer was assumed to be nonlimiting in the crop model. Other controllable variables included the latitude of the planting and the following soil characteristics: profile, albedo, water drainage constant, the lower limit of plant-extractable water, the drained upper limit of soil water, and the soil saturated water content. A root growth weighting factor was put in for each soil layer for use in distributing new root growth as the season progressed.

System parameters included the genetic characteristics and known growth patterns of each crop. Genetic coefficients for corn included the thermal time required for emergence to end of juvenile stage, the rate of photoinduction in degree-days per hour, the thermal time required for grain filling, potential kernel number, and maximum daily rate of kernel fill in milligrams per kernel. For wheat, genetic coefficients included photoperiod sensitivity, duration of grain filling, conversion of biomass to grain number and grain filling, vernalization, stem size, tillering habit, and cold hardiness. The SOYGRO model used coefficients for growth processes including photosynthesis and respiration of various plant parts; effects of temperature on development, photosynthesis, and seed growth rates; the effect of solar radiation on photosynthesis; the effect of leaf nitrogen content on photosynthesis; and the effect of physiological development on allocation of photosynthates. Genetic coefficients included the sensitivity of a cultivar to photoperiod and thermal time periods at each stage of development, maximum seed growth rate, leaf size, and the maximum rate of flower and pod addition. Many of these factors, particularly the photoperiod sensitivity, determined the latitude range for which each cultivar was adapted.

Model limitations were that (1) weeds, diseases, and insects were controlled and had no effect; (2) all nutrients were nonlimiting; (3) there were no highly problematic soil conditions such as salinity, acidity, or toxic elements, or mineral deficiencies; (4) there were no catastrophic weather events such as hail, tornados, floods, or excessive rain; and (5) the direct effects of CO_2 fertilization were not considered, except where mentioned in particular.

Climate Change Scenarios

The EPA specified that the climate scenarios used for the crop model experiments must be outputs from the Goddard Institute of Space Sciences (GISS) and Geophysical Fluid Dynamics Laboratory (GFDL). These GCMs calculated the projected climate change for steady-state $2 \times CO_2$ conditions as referenced against a baseline climate average for the years 1951 to 1980. The baseline weather data for each crop station within a GCM grid box was multiplied by the percentage changes between the model values for current averages in that grid box to predicted future averages. The current climate variability was used for the variability of future climates. Rainfall frequency was not changed with the warming temperatures. A change in rainfall frequency could have a considerable impact on plant stress. The average projected temperature rise over the United States was calculated as 4.3°C by GISS and 5.1°C by GFDL. The two models agreed that the annual precipitation over the United States will increase. The GISS model showed precipitation increasing in all four seasons, while the GFDL model had a decline in summer rainfall. (See Figure 6 in Chapter 5 for the seasonal changes in temperature and precipitation projected by the models for the United States as a whole.) The forest model calculations discussed in Chapter 5 were centered on distinctive biomes, such as the boreal, transition, and eastern hardwood forests. The crop model calculations centered on specific growing regions—the Great Lakes region for corn and soybeans, the central and southern Great Plains region for wheat and corn, and the southeastern states for soybeans and corn. The grid box sizes, in latitude and longitude, were 7.8° × 10° for GISS, and 4.4° × 7.5° for GFDL. The temperature and precipitation changes calculated for the $2 \times CO_2$ climate within each grid box were applied to all crop stations within that grid box. The details of the climate change scenarios are given below with the crop modeling results for each region.

Crop Model Results

The Great Lakes Region

Eighteen sites within this ten-state region were selected by Ritchie et al. (1989) for projecting the effect of climate change on corn and soybeans. Eight of the states (Minnesota, Wisconsin, Illinois, Indiana,

Michigan, Ohio, New York, and Pennsylvania) border the Great Lakes, and two others (Missouri and Iowa) were also included in the region. This region extends to the northern limit of the major growing region for corn and soybeans in North America. It has a short growing season due to its long, cold winters. Corn production is high in the region, but climate variability leads to considerable variation in yields. (Corn and soybeans are both warm-season crops, unlike some varieties of wheat). Most of the soils in the region were classified as medium or deep and sandy or silty loams. The crop variety selected for each site was the one best suited genetically to the current climate, but not to the projected new climate. Both rain-fed and irrigated conditions were simulated. Irrigated yields were simulated in order to assess the influence of temperature change alone on yields and to estimate irrigation needs for the future.

Climate change   The projected changes in temperature and precipitation under the GISS $2 \times CO_2$ scenario were: March to May, 4.5°C and 5.5 mm/mo; June to August, 3.5°C and 3.4 mm/mo; and September to November, 4.3°C and −14.4 mm/mo. Since the GFDL grid boxes were smaller than the GISS grid boxes, it was possible to slice the climate into three latitude belts and calculate the average climate change for the three east–west sections. These are given in Table 1.

It is clear that the growing season climate change projected by the GFDL GCM was much more extreme than that calculated by the GISS GCM. Thus one might expect the impact on crop yield to be more extreme under the GFDL scenario.

Corn yields   Calculations were made for both irrigated crop yields and nonirrigated, or rain-fed, yields. Clearly the irrigated crop yields should not be affected by changes in the rainfall amounts, provided irrigation is maintained properly. The reponse of irrigated corn yields to temperature changes was negative at all stations except Duluth, Minnesota— the northernmost station. Under the GISS $2 \times CO_2$ model climate, mean yields decreased an average of 11% (ranging from −3% at Green Bay, Wisconsin to −28% at Springfield, Illinois). At Duluth the corn yield increased 86%. Under the more drastic temperature increases projected by the GFDL $2 \times CO_2$ model, irrigated corn yields decreased an average of 43% except for Duluth, where a 36% increase was projected. Rain-fed corn yields also decreased throughout the Great Lakes region except in the northernmost parts. Under the GISS scenario, the simulations averaged −16% for most sites, with the increase at Duluth being about 49%. Under the GFDL scenario, rain-fed corn yields decreased an average of 50% with no increase at any northern latitude. This reduction

TABLE 1   Projected summer temperature and precipitation changes by latitude belt (GFDL 2 × CO₂ climate scenario)

	Degrees north latitude					
	44–49		40–44		36–40	
	Temp ($^{\circ}$C)	Precip (mm)	Temp ($^{\circ}$C)	Precip (mm)	Temp ($^{\circ}$C)	Precip (mm)
May	3.6	8.9	3.6	9.4	2.4	12.5
June	9.4	−45.0	7.0	−9.2	6.4	−6.6
July	9.4	−30.8	8.0	−39.6	7.4	−10.7
August	8.1	−13.5	4.4	0.8	4.2	7.7

(From Ritchie et al., 1989.)

was largely produced by decreases in rainfall and, in part, by a shorter growing period resulting from warmer temperatures. The extra irrigation demands of corn over the Great Lakes region under the GISS scenario was rather small (4.4%), and was probably the result of increased growing season precipitation and a decreased length of the total growing season. Irrigation water requirements for corn under the GFDL scenario were much higher due to the large decrease in summer rainfall in that model; percentage increases in irrigation averaged 50%, but ranged from −7.4% at Indianapolis, Indiana to a 174% increase at Duluth.

Soybean yields   Soybean yields using irrigation changed much less than did the irrigated corn yields. Soybean yields changed little in the GISS 2×CO₂ simulations, although there was an increase of 181% at Duluth. The GFDL 2×CO₂ calculations for irrigated soybean yields showed decreases averaging 14% except at Duluth, which had a 175% increase, and Buffalo, New York, which had a 3% increase. Rain-fed soybean yields under the GISS scenario were reduced an average of 13% for two-thirds of the locations, while the other one-third of the sites had increases ranging from 0.1% at Des Moines, Iowa to 118% at Duluth. The GFDL model calculations showed rain-fed soybean yields decreasing about 55% throughout the region, but increasing by 6% at Duluth. The irrigation water demand for soybeans under the GISS scenario increased by 10% to 40%, except at Springfield, which had a reduction of 4%. The warmer, drier GFDL scenario resulted in a 90% increased average irrigation water demand throughout the Great Lakes region.

Direct effects of CO_2   The simulations by Ritchie and colleagues were repeated to take into account the direct effects of CO_2 enrichment on corn and soybean yields. The atmospheric CO_2 concentration was assumed to increase from around 330 to 660 ppm in these calculations. CO_2 fertilization effects appeared to be more significant for soybeans than for corn, and for both crops it lessened the impacts of climate change alone. The direct effect of CO_2 increase on irrigated corn yields was less under the GISS scenario than under the GFDL model. The rain-fed corn yield was higher under the direct effect of increased CO_2 because of reduced transpiration and a better water supply. The northernmost sites had more positive yield gains than sites in the southern Great Lakes region. In general, the direct effect of CO_2 enrichment was to reduce the demand for irrigation water by corn.

Yields increased over the baseline yields for irrigated and rain-fed soybeans under the GISS scenario and for irrigated soybeans under the GFDL scenario when CO_2 enrichment was included. Yield increases ranged from 12% to 465%, with the greater increases in the north where the growing season is more limited. For rain-fed soybeans, the direct effect of CO_2 enrichment was not strong enough to overcome the yield decreases caused by the climate change. Under the GISS scenario, the direct CO_2 effect reduced soybean irrigation demand relative to that under climate change alone, but still resulted in an increased water need from the baseline climate situation. Using the GFDL scenario, the direct CO_2 fertilization effect slightly reduced water demand from that under the climate change alone in the south, and increased it in the north. In both cases it represented an increased irrigation demand (40%–200%) over the baseline climate.

The Southeastern United States

Nineteen sites within this 11-state region were selected by Peart et al. (1989) for studying the effects of climate change on corn and soybean yields. The southeastern region includes the states of Louisiana, Arkansas, Kentucky, Virginia, Florida, Alabama, Mississippi, Tennessee, Georgia, North Carolina, and South Carolina. This large region was subdivided into three areas, called Delta, Uplands, and Coastal Plains, within which the soils and climate have similarities. The region contains the effective southern limit for corn and soybean production. Soils are quite variable and many have low water-holding capacity. Most of the soils are medium silt loam and a few medium sandy loam. The climate is characterized by high summer temperatures and high precipitation variability, which places considerable stress on both soybean and corn crops. Soybeans in particular have been adapted to most of the region. Nevertheless, soybean yield is strongly affected by both rainfall and temperature occurring at the time of reproductive fruit growth. Too much water at planting time or at harvest may significantly reduce soybean yield.

Climate change   The climate change in the southeastern states was calculated using the GISS $2 \times CO_2$ model and the GFDL $2 \times CO_2$ model, as was done for the Great Lakes region. Again, the changes projected by GFDL were more extreme than those projected by GISS. The two grid boxes used for the GISS GCM were centered at Charlotte, North Carolina and Memphis, Tennessee. The eight smaller grid boxes used for the GFDL GCM were centered at St. Louis, Missouri; Greenville, Mississippi; New Orleans, Louisiana; Huntington, West Virginia; Augusta, Georgia; Gainesville, Florida; Washington, D.C.; and in the Atlantic Ocean off the Virginia coast. The calculations were done on a daily basis from the 30-year (1951–1980) weather record at each site as projected into the $2 \times CO_2$ scenarios. Typical growing season temperature increases over the 30-year baseline averages were 2.5°C and 5°C for GISS and GFDL scenarios. Under the GISS warming, precipitation increased 20% throughout most of the region, but under the GFDL warming, precipitation decreased by 50% or more. The variability of the precipitation from month to month in both GISS and GFDL scenarios was much greater than the variability of the temperature.

Corn yields   Corn yields decreased at all locations under both of the GCM $2 \times CO_2$ scenarios without the inclusion of any direct CO_2 effects on photosynthesis. Under the GISS scenario the reductions in rain-fed corn yields averaged 7.9% and under the GFDL scenario, 65%; the irrigated corn yields decreased 18% and 28%, respectively.

Soybean yields   In the Delta area, where normal soybean yields are lower than elsewhere in the southeast, the rain-fed yields were reduced about 50% under the GISS scenario and to near disaster under the GFDL case. In the Coastal Plains locations the reductions in rain-fed soybean yields were about 20% under the GISS climate and 70% with the GFDL climate. The GFDL results were more variable from location to location, with reductions of from 78% to 90%, which would represent 30 years of crop failures. There were a few locations with lesser reductions within the Coastal Plains area. Most of the Upland locations had more moderate reductions in rain-fed soybean yields than did the two other areas for the scenario, while with the GFDL climate all the Uplands reductions were near disasters. The irrigated soybean yields showed less drastic reductions than the rain-fed yields under both scenarios, since the irrigation cancelled out the reduced rainfall accompanying the climate warming. Comparison of the GFDL results for rain-fed and irrigated crops suggests that rainfall reductions were a major factor in the reduced yields.

CO_2 enrichment and soybean yields   The SOYGRO simulation model was then modified to include the direct CO_2 effects on photosynthesis and stomatal diffusion resistance to transpiration. The GISS $2 \times CO_2$ climate change scenario, which without the direct effects of CO_2 enrichment reduced soybean yields at all stations simulated, now increased yields at 12 of the 19 stations. At all other stations the yield reductions were substantially less with CO_2 enrichment than without it. The Delta area had yield reductions of about 15%, the Coastal Plains increases of about 10%, and the Upland area increases of about 20%. The average change overall was an increase of 9.1%. With irrigation, under the GISS change, yields increased with CO_2 enrichment at half the stations and the overall average yield increased 13.7%.

The GFDL $2 \times CO_2$ climate change scenario with the direct effects of CO_2 enrichment added resulted in greatly reduced rain-fed soybean yields compared with the baseline climate yields, but the reductions were somewhat less severe than those under climate change alone. The average reduction was now 54.6%, and all stations suffered equally. With irrigation, the GFDL scenario with CO_2 enrichment produced increased soybean yields at all stations, with an overall average increase of 14.9%.

Water use efficiency   Water use efficiency (WUE), as defined here, is the ratio of seed yield to total evapotranspiration during the growing season. For soybeans, there is a very strong relationship between WUE and yield. Changes in WUE are strongly related to changes in photosynthetic rates and only weakly related to changes in transpiration rates, given CO_2 concentration changes at a constant temperature. Increasing air and plant temperatures decrease WUE since they will increase the the leaf water potential and the transpiration rate. Growth chamber experiments have shown that a temperature increase from 28°C to 35°C reduces soybean WUE about 26%. If increased atmospheric CO_2 concentration increased the size of leaves and the leaf area index, reduced WUE would result. The weather, the soil, and crop characteristics all can have an effect on the evapotranspiration rate. In the GISS simulations, CO_2 enrichment significantly improved WUE since it increased yields and reduced evapotranspiration. In the GFDL simulations, CO_2 enrichment reduced WUE because it greatly reduced soybean yields. Irrigation requirements for soybeans increased an average of 33% under the GISS scenario and 133% under the GFDL simulations when the direct effects CO_2 enrichment were added.

The Central and Southern Great Plains

The central and southern Great Plains of the United States is one of the world's most productive grain regions. It includes the states of Nebraska, Kansas, Oklahoma, and Texas. The climate of the region may be characterized as highly variable and windy, with hot, dry summers and cold winters. The four states produce one-third of the nation's wheat and one-seventh of the nation's corn, primarily on deep prairie soils. Agriculture in the Great Plains is mostly dryland farming, although irrigation is used in some parts of the region. This means that agricultural production throughout much of the region is very vulnerable to droughts. This vulnerability remains in spite of the adoption of conservation tillage techniques, drought-resistant crop strains, and other improved management practices. A warmer climate may bring with it increased frequency and severity of droughts along with a lack of adequate food production and tragic economic consequences. Irrigation demand would increase, and even today there the Ogallala Aquifer, which underlies much of the western, drier parts of the region, is being overexploited and is in danger of depletion. Rosenzweig (1989, 1990) selected 14 sites within this four-state region to model the effects of climate change on wheat and corn yields.

Climate change   Climate change projections were run using the GISS $2 \times CO_2$ and GFDL $2 \times CO_2$ scenarios for the 14 sites. Because wheat is grown at all times of the year it was necessary to consider the climate change for each month. Tables 2 and 3 give the projected temperature and precipitation changes for a warmer, $2 \times CO_2$ climate. Again, the

TABLE 2   Temperature and precipitation changes with climate warming as a function of latitude and season (GISS 2 × CO₂ scenario)

	Degrees north latitude					
	39.1–47.0		31.3–39.1		23.5–31.3	
	Temp (°C)	Precip (mm/mo)	Temp (°C)	Precip (mm/mo)	Temp (°C)	Precip (mm/mo)
D,J,F	5.8	5.1	4.9	−5.4	4.7	−16.5
M,A,M	4.8	20.7	4.3	−18.9	3.7	9.0
J,J,A	3.8	0.6	4.1	−6.6	4.5	−10.2
S,O,N	5.2	−6.0	5.0	0.0	4.1	−12.0
Annual	4.9	5.1	4.6	−7.7	4.3	−7.4

(From Rosenzweig, 1989.)

TABLE 3   **Temperature and precipitation changes with climate warming as a function of latitude and season (GFDL 2 × CO₂ scenario)**

	Degrees north latitude							
	40.0–44.4		35.6–40.0		31.1–35.6		26.7–31.1	
	Temp (°C)	Precip (mm/mo)	Temp (°C)	Precip (mm/mo)	Temp (°C)	Precip (mm/mo)	Temp (°C)	Precip (mm/mo)
D,J,F	5.0	4.2	5.1	3.9	4.8	3.6	4.2	−7.4
M,A,M	4.8	7.7	5.2	9.0	5.1	−7.8	4.5	−7.9
J,J,A	7.7	−28.2	5.9	−30.1	3.3	11.8	3.2	66.4
S,O,N	5.5	1.9	4.8	3.6	4.6	−9.7	4.6	3.2
Annual	5.8	−3.6	5.3	−3.4	4.5	−0.5	4.1	13.6

(From Rosenzweig, 1989.)

GISS grid boxes (Table 2) are considerably larger than the GFDL grid boxes (Table 3).

As was done for the other regions, the baseline climate was taken from each site's weather record over the years 1951–1980. Then the climate change scenarios were developed from average monthly changes in temperature, precipitation, and solar radiation calculated for each GCM grid box for current and doubled CO_2 conditions. Observed daily climate variables were multiplied by monthly ratios of climate variables from the GCM $2 \times CO_2$ simulations over those of the simulations for the baseline conditions. Rosenzweig compared the magnitude of the climate changes using the GCM $2 \times CO_2$ scenarios with the climate of the 1930s drought years in Nebraska and Kansas. The climate change scenario temperatures were about 3°C higher than the Dust Bowl temperatures, but the precipitation decreases were about the same.

Corn yields   Although the dryland CERES-Maize model projected yields to decrease everywhere, the yield changes were significant (twice the standard deviation) at only 7 of the 14 sites when using the GISS climate projections. Yield decreases were from 4% to 43%. The mean decrease was 17%. Corn yield decreases were somewhat less at lower latitudes. This may reflect the use of higher-temperature cultivars being used in the southern part of the region. Maturity dates of dryland corn advanced between 11 and 30 days. When planting dates in the CERES-Maize were set earlier, dryland yield decreases were mitigated somewhat in some locations, but declines were still very large (up to 32%) in most locations under the GISS scenario.

Under the warmer and drier GFDL scenario, dryland corn yield decreases were very large everywhere, but particularly at the higher latitudes. Yield decreases ranged from 9% to 90%, with a mean decrease of 50%. The decreases were caused by the combined effects of high temperatures shortening the grain filling period and increased moisture stress. As seen in Table 3, the GFDL climate had a reduction of as much as 30 mm/mo in summer precipitation within the two northernmost grid boxes. This occurs during the critical growth stages of corn, i.e., flowering and grain filling. Maturity dates of dryland corn advanced by an average of three weeks.

Irrigated corn yields decreased significantly everywhere under both the GISS and GFDL scenarios, although the decreases were somewhat less than with dryland farming. Yield decreases were from 9% to 21% with the GISS scenario and from 13% to 37% with the GFDL scenario. The more extreme GFDL climate produced greatly increased irrigation water demand at all sites, while the GISS simulated climate increased irrigation water demand at only half the sites.

CO₂ enrichment   When the direct effects of CO_2 enrichment on dryland corn were included in the GISS scenario, the yield increased at all sites. Under the GFDL scenario and CO_2 enrichment there was a decrease of yield at all sites, but it was somewhat less than it was without the fertilization effect. Irrigated corn yields decreased compared with the baseline yields at all sites under both GCM scenarios when the direct effects of CO_2 enrichment were included, but the reductions were not as great as they were without fertilization. The decreases occurred despite the effect of increased photosynthesis combined with decreased stomatal conductance. This simulated response by CERES-Maize is caused by the high temperature advancement of developing stages such as grain filling, which produced yield decreases even though photosynthate production increased and water use efficiency increased.

Water use efficiency   WUE increased everywhere under the GISS scenario and at half the sites with the GFDL scenario when CO_2 enrichment was included. The increases in WUE reduced the demand for irrigation water everywhere under the GISS scenario with direct CO_2 effects. These changes were caused by decreased stomatal conductance and the shortening of the growing season. But under the GFDL scenario, these two factors were unable to overcome the hotter and drier simulated climate, and water demand for irrigation increased significantly over the baseline climate water demand in the northern and central portions of the region.

A Shift of the Corn Belt

A strictly empirical method was used by Blasing and Solomon (1983) to project the potential shift of the corn belt in the United States with

FIGURE 1 **The bold solid line outlines the North American corn belt; the numbers correspond to geographic locations whose characteristics are described in Figure 2. (From Blasing and Solomon, 1983.)**

climate warming. Their first step was to evaluate the heat and moisture requirements of corn within the present-day corn belt. The thermal requirements of corn can be expressed as heat sums in the form of growing degree-days (GDD); i.e., the mean daily temperature accumulated over all days when it exceeds 10°C. (Earlier, in Chapters 4 and 5, we used GDD above a 5°C threshold. The threshold depends upon the particular plant being considered.) Moisture requrements for corn were also considered in this study. Annual precipitation is an important parameter for corn because of its relationship to the amount of stored soil moisture available for the rapid growth that takes place in July. The distribution of precipitation during the summer months is also important, but Blasing and Solomon found that precipitation during the summer months could be estimated to a reasonable approximation from annual precipitation.

Figure 1 shows the geographic distribution of the corn belt and the climate divisions used for its characterization. Regionally averaged temperature and precipitation data for the United States are available from the National Climatic Center in Asheville, North Carolina. Each state is divided into from one to ten regions, known as state climatic divisions. Blasing and Solomon obtained averages of annual precipitation for the period 1941–1970 for each climatic division in or partially within the

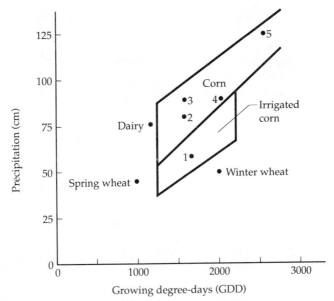

FIGURE 2　Heat and moisture characteristics of the North American corn belt. The solid lines represent the climate niche of corn. The numbers 1 to 5 correspond to the geographical locations in Figure 1. Points outside the lines correspond to dairy production in central Wisconsin, spring wheat in central North Dakota, and winter wheat in west-central Kansas. (From Blasing and Solomon, 1983.)

corn belt, as well as annual degree-day values computed from monthly data. They then plotted those annual precipitation and GDD values as shown in Figure 2. These plotted points all fell within the parameters represented by the solid lines in Figure 2. One exception is represented by point 5 for western Tennessee, which falls outside of what is considered to be the corn belt. However, corn is grown there in places where soil and topographic conditions permit. According to Blasing and Solomon:

This suggests that climate is not the limiting factor for corn production, so such regions were included in this characterization of the climatic requirements for corn. However, corn-producing regions near the Gulf Coast were not included. To extend those climatic conditions northward to places where soil and topographic conditions once again favor corn as the major crop would require climatic changes beyond the scope of this study.

For a given amount of annual precipitation, the climate may become too warm to favor corn production, and winter wheat may predominate instead. Winter wheat is a major crop in southeastern Kansas, characterized by 2400 GDD, but corn is the more important crop in northwestern Illinois, characterized by 1600 GDD, even though the annual amounts and seasonal distributions of precipitation are nearly the same in those two locations. Northwestern Illinois has less heat stress and lower annual evapotranspiration than does southeastern Kansas. A climate that today is too warm to be in the corn belt may fall within it when the climate gets warmer if the annual precipitation increases sufficiently. Blasing and Solomon pointed out that the distinction between dryland and irrigated corn in this study is arbitrary and is based on current practice within the western part of the corn belt.

Blasing and Solomon used the climate change projected by Manabe and Stouffer (1980) for the $2 \times CO_2$ climate. This GCM projection gave a global average temperature increase of 2°C, with a summer and winter increase of about 3°C in the corn belt. GDD increases were obtained by adding 3°C to each monthly temperature and recalculating the sums for the year. Precipitation increases were estimated for the corn belt using $2 \times CO_2$ climate projections computed by Manabe et al. (1981); these GCM results were earlier versions of the climate simulation referred to above as GFDL $2 \times CO_2$. They projected a precipitation increase for a latitude around 45° N, which was used here since it coincided with the present-day corn belt.

The parameters in Figure 2 were used to identify those regions expected to have heat and moisture conditions characteristic of the present-day corn belt under four different climate change scenarios. The first projection was a geographic shift of the corn belt with a 3°C increase in temperature over current GDD conditions and with no change in precipitation (Figure 3A). The corn belt would move into southern Canada, although not far. The northeastern part of Minnesota would still be too cool for good corn production. A large section of the present-day corn belt in Kansas, Nebraska, South Dakota, Missouri, Iowa, and southern Illinois would become unsuitable for good production, even of irrigated corn.

The second projection of a corn belt shift with climate change used the same 3°C warming, but added 8 cm to the annual precipitation. This climate reduced the area suitable for irrigated corn, but expanded the projected corn belt slightly to the south and to the west. The upper peninsula of Michigan now became too wet, probably the result of too much moisture for planting in the spring and too cool a soil. The projected northward movement of the corn belt was least in the Great Lakes region. Apparently, the moderating effect of these large bodies of water has a significant regional influence. Along the Canadian border

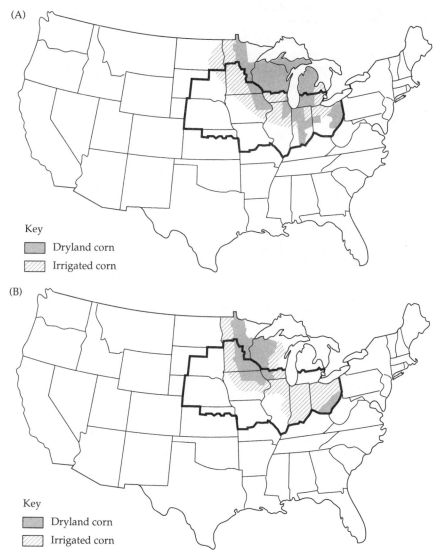

FIGURE 3   (A) Geographic shift of the corn belt projected with a 3°C temperature increase evenly distributed over the year and with no change in precipitation. The bold solid line bounds the current corn belt. (B) Geographic shift of the corn belt projected for a 3°C temperature increase evenly distributed over the year and an 8 cm increase in annual precipitation, but with July–August precipitation 0.6 cm less than current values. (From Blasing and Solomon, 1983.)

the GDD requirement for corn is close to the minimum thermal sum necessary for its growth.

Although the annual precipitation increased in the model of Manabe et al. (1981), the July precipitation showed a decrease of 0.6 cm under the $2 \times CO_2$ climate. Dryland corn requires 17 cm or more of precipitation during July and August (19 cm if the average July temperature exceeds 27.7°C) and irrigated corn 15 cm (16.5 cm if the July temperature exceeds 27.7°C). The corn belt shift was calculated for the $2 \times CO_2$ climate using a 3°C temperature increase and an 8 cm increase in annual precipitation, but with the July and August precipitation 0.6 cm less than the current average value, and with the requirement of 1.5 to 2 cm additional precipitation during the July–August period if the mean July temperature exceeded 26.5°C. Now the changes were great, as seen in Figure 3B. All of Michigan was outside the corn belt, and most of the states of Illinois, Indiana, and Ohio required irrigated corn, as did Iowa, Minnesota, and southern Wisconsin.

Finally, a fourth projection was made for the $2 \times CO_2$ climate based on the premise that increased temperatures would lead to a longer growing season and therefore to earlier possible planting dates. Corn could then reach the silking–tasseling stages earlier and precipitation in June and July would become more critical. For this fourth projection, Blasing and Solomon decreased the July precipitation 0.6 cm and held the June precipitation constant at its current value. The average temperature increase again was 3°C. The July–August moisture requirement for corn was now applied to the June–July period. The main result of this projection was that dryland corn replaced irrigated corn throughout the area from southern Wisconsin through Illinois and Ohio.

These projections were considered to represent extreme cases by the authors of this study. They admitted that current GCM projections are not sufficiently refined for a specific region to yield results that are likely to be realistic with a high degree of probability. They also did not take into consideration either increased water use efficiency with CO_2 enrichment or crop breeding potential. However, these results are not inconsistent with model projections for corn production by other researchers, including Newman (1980), who found the corn belt to shift along a SSW to NNE direction by 175 km per °C, whether with warming or cooling.

Wheat Production in North America

Wheat is the world's most important and most extensively planted grain crop and is produced in greater amounts than any other crop. In the United States, wheat ranks second to corn in acres planted and is third in export value after corn and soybeans. Rosenzweig (1985) investigated the influence of the expected $2 \times CO_2$ climate change on the areal extent of the wheat-growing regions of North America, identified areas particularly vulnerable to climate change, and finally examined the Southern Great Plains region in greater detail.

Wheat is grown in a wide range of climates and soils. Wheat cultivars vary greatly in their responses to environmental conditions. Wheat is classified according to its growth habit (winter, spring, or fall-sown spring) and its kernal texture (hard or soft). The major wheat-growing areas of North America are shown in Figure 4A. Rosenzweig used the GISS GCM to simulate the current climate and the $2 \times CO_2$ climate. (At the time of her work, the GISS GCM used slightly larger grid boxes than did the later model referred to earlier in this chapter.) She then identified the types of wheat best suited to each of the regions defined by the grid boxes covering the appropriate parts of North America; these are shown in Figure 4B. Next she simulated the North American wheat regions using the GISS current-climate control run (average annual temperature and precipitation 1930–1960); the results of this simulation are shown in Figure 4C. Of 32 grid boxes, the model reproduced all but 7 correctly in both growth habit and kernel texture. (The main problem with getting a more precise agreement is the coarse nature of the GISS grid and some of the management practices in specific areas.)

Environmental Requirements of Wheat

The environmental requirements for wheat, according to its growth habit and kernel texture classification, are shown in Table 4. Different cultivars of wheat vary with respect to the time they take to develop. Some spring wheat varieties mature in 75 to 95 days, while others require 95 to 115 days. Very little wheat is grown where the frost-free period is under 90 days because of the likelihood of a freeze. Rosenzweig used 90 days as the minimum growing season length. GDD were calculated on a monthly basis from mean monthly air temperature sums over the growing season. The temperature contribution was 0 when the value fell below 4°C or above 32°C. The demarcation line dividing winter wheat areas from spring wheat areas runs close to the -13°C mean minimum January isotherm. Rosenzweig selected the -12°C isotherm to separate winter from spring wheat production. Winter wheat has a strong vernalization requirement (a cold period is necessary to promote floral induction); spring wheat does not. Vernalization of winter wheat occurs between 0°C and 11°C, with an optimum around 3°C for a duration of 6 to 8 weeks. Rosenzweig used a vernalization requirement of 5°C for a minimum duration of one month. If the mean monthly air temperature did not fall below 5°C at any time during the year, then

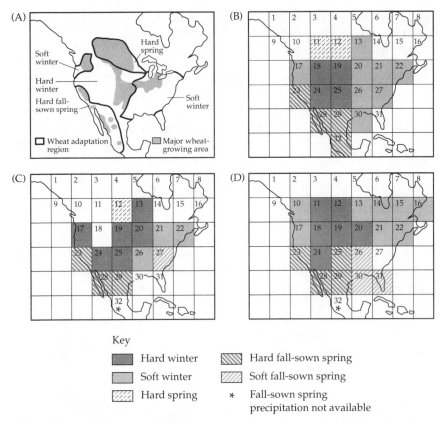

Key

■ Hard winter   ▨ Hard fall-sown spring

▦ Soft winter   ▨ Soft fall-sown spring

▨ Hard spring   *   Fall-sown spring
                     precipitation not available

FIGURE 4   **(A) Major wheat-growing areas of North America. (B) Actual wheat-growing regions of North America as defined by the GISS GCM grid. (C) Simulated North American wheat regions using the GISS GCM control run. (D) Simulated wheat regions using the GISS GCM 2×CO₂ climate. (From Rosenzweig, 1985.)**

the wheat type for the region was fall-sown spring wheat and not winter wheat.

Shift of the Wheat-Growing Regions

The GISS 2×CO₂ climate scenario was used to project the expected shift of the wheat-growing regions in North America with climate change. In this scenario, the global average annual surface temperature increased by 4.2°C; this is near the maximum projected by the various climate models. The model used was an earlier and more extreme version of the GISS models. Its use in this study means that the projected shift of the wheat-growing regions is a maximum. The model

TABLE 4   **Environmental requirements for wheat**

Length of growing season	90 days
GDD per growing season	1200 GDD
Mean minimum January temperature	
Spring wheat	Below −12°C
Winter wheat	Above −12°C
Vernalization requirement	
Winter wheat	At least one mean monthly temperature less than 5°C
Fall-sown spring wheat	Mean monthly temperature for all months greater than 5°C
Annual precipitation (mm/yr)	
Soft wheat	760–1200
Hard wheat	0–760
Dry moisture conditions	0–380
Adequate moisture conditions	380–760

(From Rosenzweig, 1985.)

projected temperature increases of 4.2°C for the eastern United States and 4.9°C for the central and western regions. The temperature increase was about 40% greater in winter than in summer. Precipitation changes were calculated as percentages of the observed baseline climate (1930–1960) using the GISS $2 \times CO_2$ scenario and then were applied to the actual precipitation amounts to estimate the $2 \times CO_2$ precipitation. Most of the grid boxes showed an increase in annual precipitation; only a few showed a decrease, mostly those along the Gulf Coast and in the eastern Great Lakes and New England.

In Rosenzweig's simulations, the warmer $2 \times CO_2$ climate caused a shift of the wheat-growing regions northward into Canada, while at the same time enhancing some of the productive regions within the United States (Figure 4D). Along the Canadian border, temperatures moderated with the warming so that the growing season was lengthened, the GDD increased, and the mean minimum January temperature raised. Now, no spring wheat, either hard or soft, was projected to grow north of the Canadian border, whereas some was grown there using the present baseline climate. Much of the winter wheat grown there was soft since the annual precipitation was now greater than 760 mm. Hard winter wheat in the Pacific Northwest shifted to soft winter wheat with increased precipitation, and Montana became more productive. Hard and soft fall-sown spring wheat now was projected to grow at most southern latitudes. Moderate winter temperatures also shifted winter wheat to fall-sown spring wheat in Kansas, Oklahoma, Arkansas, and Indiana. In Mexico, wheat-growing regions remained the same, but yields were projected to go down as the result of much more heat stress.

The result of the $2 \times CO_2$ climate change was generally favorable for the production of wheat throughout much of North America. The winter wheat-growing regions extended into Canada and the fall-sown spring wheat regions moved northward and eastward. Seven more grid boxes were suitable for wheat production under the $2 \times CO_2$ scenario than in the baseline control run. Rosenzweig was quick to point out that this study had a number of limitations and that the changes in temperature and precipitation shifted the wheat classification just under or over the limit into the next category. In these situations, some wheat of each type might be able to grow within a grid box. Higher mean temperatures during wheat growth may increase the need for earlier-maturing, more heat-tolerant cultivars throughout North America.

Changes in Wheat Yields

Rosenzweig (1989) used the CERES-Wheat model to estimate wheat yield changes under $2 \times CO_2$ climate warming in the central and southern Great Plains. The climate change scenarios used were the GISS and GFDL $2 \times CO_2$ climate warming models, as specified by the U.S. Environmental Protection Agency and described earlier; see Tables 2 and 3 for the projected climate changes by latitude and season for the central and southern Great Plains region.

Wheat yields simulated using the GISS $2 \times CO_2$ scenario decreased relative to those simulated by the current climate baseline scenario at every location in the central and southern parts of the Great Plains, with larger decreases toward lower latitudes. Dryland yield decreases ranged from 10% to 55%, with a mean decrease of about 30%. The yield decreases were primarily the result of the higher temperatures reducing the duration of certain growth stages, such as grain fill. Shortening of the grain-filling period reduced the amount of carbohydrate available for grain formation and reduced the yield. The GISS scenario moved the maturity dates for wheat forward by about three weeks. Also, the shorter growing season resulted in a reduction of the total seasonal evapotranspiration, even though the daily amounts were increased by the higher temperatures.

Dryland wheat yields under the GFDL scenario were similarly decreased, by from 12% to 55% from the baseline climate. The mean decrease was 33% for the entire region. Maturity dates advanced by up to four weeks using this slightly warmer scenario. The same reasons for reduced yields apply here as with the GISS example.

Irrigated wheat yields were reduced under both climate scenarios, but not as much as those of dryland farming. Also, the total growing season evapotranspiration reduction was not as great with irrigation as it was without it. Maturity dates under both scenarios were about three weeks earlier than under the baseline climate. Under the GISS scenario, water demand for irrigation remained about the same at the more northern latitudes, where precipitation increased, and water demand increased at the more central and southern latitudes where precipitation decreased a great deal during the growing season.

Rosenzweig writes:

Planting winter wheat too early in the fall can decrease yields because of excessive growth before the onset of cold weather. If global warming extends the period between last frost in the spring and first frost in the fall, farmers may adjust planting date of winter wheat accordingly. When planting date windows are delayed in dryland and irrigated CERES-Wheat simulations with the GISS $2 \times CO_2$ climate change scenario, yields improve over those for the original planting date under this scenario in only a few cases. This shows that the modeled yield decreases in the climate change scenario are not caused by too early fall planting.

When better-adapted cultivars are used, wheat yields under the $2 \times CO_2$ climate are near or above the baseline climate yields at two-thirds of the dryland sites. With irrigation, the yields are equal to or above the baseline climate yields at more than half the locations.

Carbon dioxide fertilization of wheat mitigated the climate warming effects on wheat yields at some, but not all, locations under both GCM scenarios. Sites at the more southern latitudes showed less enhancement with the CO_2 enrichment than did more northern sites. With irrigation, wheat yields improved with CO_2 fertilization over the baseline climate yields at all except the most southern locations under both climate warming scenarios.

Irrigation Water Requirements in the Great Plains

Irrigated agriculture has replaced native vegetation, low-value crops, and dryland farming in many parts of the Great Plains. The depletion of water in the Ogallala Aquifer is a threat to irrigated agriculture throughout the region. Any increase in irrigation water use due to increased evaporative demands may accelerate the depletion of ground and surface waters. Allen and Gichuki (1989) modeled the demand for irrigation water by a variety of crops growing at 17 stations in Texas, Oklahoma, Kansas, and Nebraska using the GISS and GFDL $2 \times CO_2$ climate warming scenarios. They incorpoated the direct effect of CO_2 fertilization on crop stomatal resistances by increasing them about 20%.

In these simulations, the length of the growing season increased throughout the region, with increases ranging from 0% to 28% the smallest being near the southern border and the greatest in the north. Changes in growing season lengths depended on planting dates for the various crops (see the discussion above for wheat). Major changes in irrigation water demand were projected for all 17 stations. Crops making use of longer growing seasons, such as alfalfa, demanded increases of irrigation water. These increases were generated by higher air temperatures, more wind, more solar radiation, and a longer growing season. Decreases in seasonal net irrigation requirements were projected for winter wheat and corn, particularly with the increased stomatal resistances anticipated with increasing CO_2 levels. Some decreases in irrigation water demand resulted from shortened growing seasons where accelerated crop development took place.

Allen and Gichuki wrote:

Adaptation of a longer-season variety of winter wheat having a 20% greater growing season solar radiation/degree day requirement would likely increase growing season lengths by 15 to 20 days in the Great Plains and would increase seasonal irrigation requirements by about 10% in Nebraska, Kansas, and Texas and by about 20% in Oklahoma as compared to current cultivars under the $2 \times CO_2$ settings. Thus, if farmers adapted to crop varieties with higher seasonal environmental energy requirements to more fully utilize increased levels of available solar radiation and temperature by changing to longer season crop varieties and/or increasing cropping intensity, irrigation water requirements would increase.

California Agriculture Yields and Climate Change

California produces approximately 10% of the total farm income in the United States from a wide diversity of irrigated crops. These include grapes, cotton, hay, lettuce, almonds, tomatoes, strawberries, oranges, broccoli, walnuts, sugar beets, peaches, and potatoes (listed in descending order of economic importance). Because of this crop diversity, Dudek (1989) used a simple empirical approach to estimate crop yield changes under the GCM-modeled $2 \times CO_2$ climate warming. Since all of these crops require irrigation, consideration had to be given to future water demand and availability.

Agricultural irrigation accounts for about 80% of all consumptive water use annually in California. The major hydrologic basins and their agricultural water use shares are: the San Joaquin and Tulare, 51%; the Sacramento, 25%; and the Colorado, 12%. The heaviest water users are cotton, feed grains and hay, and pasture. Crop demand for water ranges from about 7.9 acre-feet per acre of rice in the Sacramento Valley to 1.2

acre-feet per acre of wine grapes on the Central Coast. Average annual precipitation in California is about 584 mm, of which nearly 60% is evaporated and transpired by native trees, shrubs, and grasses. The state has an enormous irrigation network that captures and distributes water that falls within the state and, in addition, brings in water from Oregon streams and the Colorado River. Water supply sources by percentage of annual use are: local surface water, 27%; groundwater, 17% : groundwater overdraft, 6%; the federal Central Valley Project, 20%; State Water Project, 7%; Colorado River, 15%; and others, 8%.

In order to estimate the effects of climate change on crop production in California, Dudek used an agro-ecological zone method to match crops to specific production regions most likely to support high potential yields. Implicit here is the assumption that the maximum yield of a crop is primarily determined by its genetic characteristics and its adaptation to the environment. Specific soil properties were not considered here, nor did this analysis allow for insect pests and pathogens, or for changes in farm management practices.

California's agricultural regions are shown in Figure 5. Dudek's study used data on average monthly temperatures and average monthly cloud cover from five California weather stations. One station, Los Angeles, represented the coastal regions; interior regions were represented by Red Bluff in northern California, Sacramento, Fresno, and Blythe. In order to estimate the effects of climate changes, the baseline climate data for each weather station was multiplied by the average monthly temperature change ratios generated by the GISS and GFDL $2 \times CO_2$ projections for northern and central/southern Californ.ia. Modeled cloud cover was also used. In central and southern California, mean temperatures were projected to rise 3.8°C to 4.4°C. In the northern part of the state, the projected increases were from 4.3°C to 5.0°C.

A Simple Empirical Crop Model

Dudek's crop production model used a linear equation containing a series of empirically evaluated parameters for each specific crop. The equation included the following variables: the relationships of potential maximum yield to crop development and leaf area; dry matter production for both cool and warm conditions; total length of the growing period; fraction of the daytime the sky is clouded; maximum leaf gross dry matter production rate for a given climate; and gross dry matter production of a standard crop for a given location on a completely overcast day and on a clear day. The growing season is an important variable. Higher temperatures might result in a shorter growing season, promoting more rapid plant growth. This in turn might eliminate any overall increase in yield for certain crops. In fact, yields might even

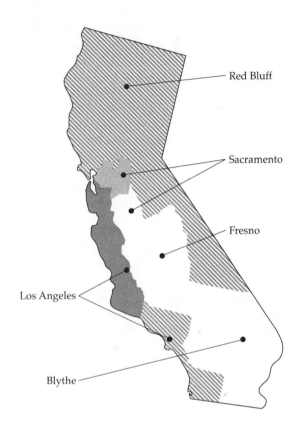

decrease under circumstances where rapid growth is triggered by early growing season warming and a shortened time for the production of harvestable yield. The maximum leaf gross dry matter production rate is a function of the average daily temperature over the growing season.

Dudek modeled the yields of four indicator crops. Each crop had a specific temperature function, in which production rate increased at low temperatures, reached a very broad optimum, and decreased at high temperatures. The optimum temperature for sugar beets was from 13°C to 22°C, for cotton and tomatoes from 20°C to 35°C, and for corn from 25°C to 35°C. Data on crop planting time and length of growing season for these four indicator crops were entered into the model. The model's projections of percentage changes in yields for the indicator crops under the $2 \times CO_2$ climate warming were then applied to other crops with similar characteristics.

It was necessary not only to model the effects of temperature changes on crop productivity, but also to attempt to estimate the impacts

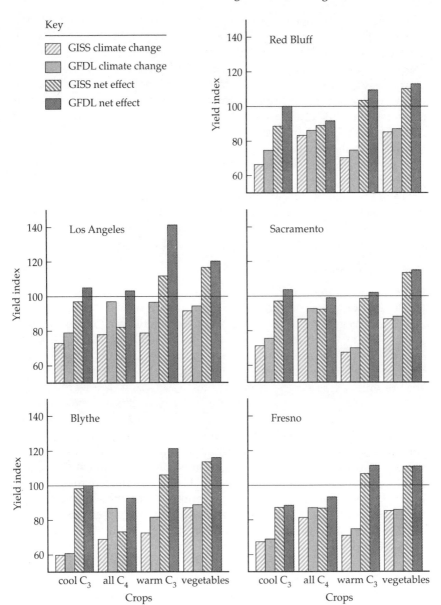

FIGURE 5 Regional productivity changes in California projected for a 2×CO₂ climate change using a GISS GCM climate scenario and a GFDL GCM scenario, under climate change alone, and with the direct effects of CO₂ enrichment (net effect). Changes relative to the baseline climate are shown for cool-season C₃ plants, warm-season C₃ plants, C₄ plants, and vegetables. (From Dudek, 1989.)

of hydrological changes that would result from climate change. Dudek used the hydrology of the Sacramento–San Joaquin basin and the work of Lettenmaier (1989) to estimate these impacts. The Sacramento–San Joaquin hydrology is strongly affected by snow accumulation in the Sierra Nevada mountains. The details of Dudek's evaluation of the effects of hydrology changes on crop production are not clear, but it appears that with the warmer $2 \times CO_2$ climate, the water available for irrigation will diminish considerably. The general warming will decrease snow accumulations in all of the catchment basins considered. Reduction in the amount of precipitation occurring as snow will increase winter runoff and a decrease of spring and summer runoff. The GCMs projected increased precipitation as rainfall in the winter months and more moisture available in the early spring, but also increased evapotranspiration in the spring and summer. However, the consensus is that the projection of future water supplies is still a highly uncertain game.

California Crop Yields

Figure 5 shows the productivity changes projected using the GISS and GFDL $2 \times CO_2$ scenarios for cool- and warm-season C_3 crops, all C_4 crops, and vegetables. Also shown are the productivity changes anticipated when CO_2 enrichment effects were added to the climate change scenarios. The cool-season C_3 crops, as calculated for sugar beets, were estimated to suffer the most serious productivity declines with climate change only; these ranged from 21% to 40%. Cotton, the warm-season C_3 indicator crop, had projected productivity declines very similar to those for corn, a C_4 crop; these declines ranged from 3% to 31%. Vegetable crops, represented by tomatoes, were the least affected by climate warming, with reductions ranging from 5% to 16%. As expected, these productivity impacts were distributed along a south-to-north gradient. With the exception of corn, yields were affected most in the interior regions of California, where temperatures are less modulated by marine influences. The GISS model produced larger temperature changes for California than did the GFDL model; therefore the GISS yield reductions were the greater of the two projections.

Effects of CO_2 enrichment When CO_2 fertilization of the crops was added to the model, many of the yield reductions projected for the $2 \times CO_2$ climate warming were mitigated; in fact, in all cases the yields increased relative to those under climate warming alone. Corn showed the greatest yield decline of all the crops under climate change alone and was the least able to take advantage of CO_2 enrichment. Cotton benefited the most from CO_2 enrichment and showed net changes from the baseline yields from a modest 1.5% decline under climate change alone to a 41% increase with CO_2 enrichment. Similar improvements were found for sugar beets.

The MINK Region

The MINK region of the central United States comprises the states of Missouri, Iowa, Nebraska, and Kansas. This region has mostly level to gently rolling terrain, except in the southeast where the Ozark Mountains reach their northernmost extent. The soils are among the most agriculturally productive in the world. There is a gradual increase in elevation from the eastern edge (70 m above mean sea level) to the western boundary (1655 m above mean sea level). Its climate is continental: winters are cold and dry and summers are hot and moist. Precipitation declines from east to west. Precipitation reaches a peak in June except in Missouri, where it peaks in May. The Rocky Mountains impede the flow of moisture into the region from the Pacific Ocean. Southerly winds from the Gulf of Mexico provide much of the summer moisture, a situation caused by the Bermuda high-pressure system that lies off the East Coast during the summer months. The growing season ranges from 1650 GDD in the south to 1310 GDD in the north. The region encompasses portions of four major U.S. land resource areas: the Central Feed Grains and Livestock Region, the East and Central Farming and Forest Region, the Central Great Plains Winter Wheat and Range Region, and the Western Great Plains and Irrigated Region. In the 1980s, the region produced 34% of the nation's corn, 30% of its soybeans and winter wheat, and 50% of its grain sorghum.

The EPIC Crop Model

The U.S. Department of Energy sponsored a study entitled "Processes for Identifying Regional Influences of and Responses to Increasing Atmospheric CO_2 and Climate Change: The MINK Poject." The results of this study were presented in a series of reports following an overview by Rosenberg (1991); each report by a different author is an appendix to Rosenberg's. To project primary crop production, the researchers used the Erosion Productivity Impact Calculator (EPIC), a semi-empirical process model that simulated crop biomass and yield, evapotranspiration, and irrigation requirement. The EPIC model was also used to calculate the direct effects of increased CO_2 concentration on photosynthesis and evapotranspiration.

Over 50 model farms distributed throughout the four states were designed for use in the model. The farms differed as to soil type, crop rotation, and management practices. When the current weather (1951–1980 means) and current CO_2 concentration were used as inputs in model simulations, the resulting crop yields and water use were found to be in reasonable agreement with actual yields and water use. Since all GCMs predict that the climate will get warmer, and maybe drier, with a doubling of the atmospheric CO_2 concentration, the researchers, rather than using a GCM climate scenario, decided to impose an historical climate event on the model representing an analogue of the predicted greenhouse-forced climate change, and to calculate crop productivity and economic responses to that event. The decade of the 1930s was chosen as the analogue of the climate change. This was the "Dust Bowl" period in the MINK region, when conditions were warm and dry. Mean annual precipitation was 28 mm less than in the control period (1951–1980) in Missouri, and in Kansas it was 102 mm less. The regional average temperature was 0.9°C higher in the 1930s than in the control period. In Missouri it was only 0.7°C higher, but in Nebraska it was nearly 1°C higher.

The analysis proceeded using four specific tasks as follows:

A. Current climate and baseline description of region
B. Imposition of climate change on the current baseline
    B1. Climate change only, no adaptation
    B2. Climate change plus 100 ppm increase in atmospheric CO_2, no adaptation
    B3. Same as B1, with currently available adaptation methods
    B4. Same as B2, with currently available adaptation methods and higher CO_2
C. Baseline description of region in the future
D. Imposition of a climate change on the future baseline and a repeat of B1 through B4, now designated D1 through D4, but using future adaptation methods in place of current methods

The EPIC Crop Simulation Results

According to the EPIC simulations, dryland crop yields were reduced an average of 20% to 25% (except for wheat) under the B scenarios. Wheat was barely affected by B1. Higher CO_2 in the B2 scenario offset yield losses because of increased photosynthesis and reduced evapotranspiration. With irrigation, all losses were reduced and were essentially eliminated under elevated CO_2 in the B2 scenario. Yields of irrigated wheat increased in the warmer, drier analogue climate. Evapotranspiration increased on irrigated land by about 25% for summer crops and by 10% for winter wheat. The currently available adaptation methods used under scenarios B3 and B4 included earlier planting, using longer-season crop varieties, and changes in tillage to increase water conservation. Yield losses were generally smaller with these adaptations, and in some cases under higher CO_2 concentration the yields increased.

The future baseline projection was for the year 2030. This study projected the growth of agriculture in the MINK region, taking into account the growth of global markets, to show an increase in crop yields of about 67% over the current baseline in the absence of climate change. Evapotranspiration decreased by about 4%. The yield increases were bred into the plants by hastening leaf development and increasing photosynthetic efficiency, improving the harvest index (the portion of the total biomass in harvestable organs), and improving pest resistance. In addition, the researchers allowed for improvements in harvest efficiency over the 45-year period.

Then the analogue climate was assumed for the year 2030 and crop yields were again calculated. All crop yields except wheat suffered reductions under the D1 scenario, and also under the D2 scenario, which included CO_2 enrichment. The percentage losses were similar to the losses under the B1 and B2 scenarios. Absolute losses were higher with the future climate scenarios, but yields were still above those of the baseline A scenario. In the D scenarios, two additional adjustments were made: reduced stomatal conductance was used as a surrogate for drought resistance, and improved irrigation efficiency was assumed. When these were applied in both the D3 and D4 scenarios yields increased considerably. The increased CO_2 concentration plus the adjustments raised all crop yields above those in scenario C (future baseline without climate change) except for corn, which was only 7% below that yield. The yields of cool-season and irrigated crops benefited more than did warm-season crops. In the worst-case D scenario (no CO_2 enrichment and no additional adjustments), production, and essentially yields, decline 22% from the baseline projection for the year 2030. In the best-case scenario (CO_2 enrichment and additional adjustments), the yields of sorghum, wheat, and soybeans increased, although corn declined, and total production of the four crops increased 3% from the 2030 baseline scenario.

The MINK study concentrated considerably on the economics of production, a subject I have elected not to include in this discussion. These scenarios for crop yields in a warmer and drier MINK region are certainly far from disaster; but this is only the projected situation for the year 2030, not 2070 and beyond. These yield reductions are considerably less severe than those in the GCM $2\times CO_2$ scenarios used in the EPA-sponsored analyses. There the temperature increases expected were closer to 3°C or more, rather than the 0.9°C as used here.

Another study of soybean, wheat, and corn yields with the greenhouse warming over the Great Plains was presented by Okamoto et al. (1991). These authors used more extreme climate changes than most of the other studies, but their results were similar to the projections given above, although different in detail.

Climate Impacts on Pest–Plant Interactions

Crop pests are defined as weeds, insects, or disease pathogens. Agriculture involves three or more trophic levels—for example, the plant, the pest, and the predator or pathogen. Climate affects all trophic levels but may affect one level more than another. Temperature increases, even small ones of 1°C to 3°C, may strongly affect the growth rates of many pests, such as bacteria, fungi, nematodes, weeds, and insects. All life stages may be affected. Virus replication is highly temperature dependent. Droughts may enhance the growth of sor pest populations by altering the quality of their food supply, or in ibit other pest populations by creating less than optimum conditions or their reproduction and survival. Temperature and humidity, either too high or too low, may acutely affect fecundity and larval survival in many insect populations. Mexican bean beetles, a pest of soybeans and other leguminous crops, are greatly reduced in numbers by moderately high temperatures and low humidity. On the other hand, spider mites, a pest of soybeans, apples, corn, sorghum, wheat, and tomatoes, are favored by warmer temperatures but reduced in numbers by high humidity. Most airborne pathogenic fungi require nearly 100% relative humidities for germination and infection.

Males and females of a given insect species may be differentially affected by climate, and this will skew the sex ratio of the population. Warmer temperatures often increase the activity levels of an insect. Ladybird beetles, a predator of aphids, greatly increase their activity rates with increased temperature. Aphids are pests of corn, cotton, alfalfa, peas, hops, and potatoes. Lower temperatures during cool weather allow aphids to grow rapidly without the deterent effect of predators. Reviews of the major insect pests and pathogens with respect to climate and environmental stresses are found in Stinner et al. (1989) and Porter et al. (1991).

Migratory Insects and Climate Change

The U.S. Environmental Protection Agency supported a study by Stinner and his colleagues (1989) which modeled the effects of climate change on insect–crop interactions. They studied five migratory insect species: the potato leafhopper, the green cloverworm, the sunflower moth, the black cutworm, and the corn earworm, although they modeled only the corn earworm in any detail. They predicted that warmer winter temperatures will have significant effects on the populations of many temperature-limited insect pests. This is especially true of migratory insects that overwinter in milder regions, such as the Gulf Coast and then move northward for the summer. Stinner and colleagues considered the overwintering ranges of four of these migratory insect pests under the current climate and then projected the changes of these ranges under the GISS and GFDL $2\times CO_2$ climate scenarios.

The potato leafhopper is a serious pest on many crops, including alfalfa and soybeans. Under present climate conditions, this pest overwinters within a narrow band along the coast of the Gulf of Mexico. Increases of winter temperatures, as projected by both climate change scenarios, imply a doubling or tripling of the overwintering range with the warmer $2\times CO_2$ climate. Increases of the overwintering population of potato leafhoppers by this amount would increase the invading populations in the northern states proportionately. The invasions would also be earlier in the growing season, and both factors would lead to increased crop damage.

The green cloverworm, sunflower moth, and black cutworm have much more extensive overwintering ranges, so the effects of climate warming on those species should not be as significant as with the potato leafhopper. Black cutworms damage corn in northern midwestern states, and the green cloverworm is a potential threat to soybeans. With climate warming, each will become a much more serious pest to these important economic crops.

The corn earworm The corn earworm feeds on many crops, including corn and soybeans, upon which very large populations develop. Populations of this insect can vary greatly from field to field and from year to year, thereby requiring models which are year- and site-specific. The corn earworm is a migratory pest that overwinters throughout most of its range in the pupal stage. Pupae breaking diapause develop into adults that lay their eggs on corn. Later adult populations then invade other crops, including soybeans. For example, the first adults are found on corn in June and July, but larvae are not found on soybeans until early August. Currently, the corn earworm is not a serious pest to soybeans in the Midwest, but it is troublesome to corn. In the South, the corn earworm goes through numerous generations and does damage to both corn and soybeans. It then migrates northward and damages corn, but does not have time to develop on soybeans. Stinner and colleagues modeled the response of the corn earworm on soybeans to climate change using the GISS and GFDL $2\times CO_2$ scenarios. They used two submodels, one for soybean response to climate change and one for the corn earworm.

The soybean submodel Soybeans have several distinct stages of development. Six out of eleven stages depend on both night length and temperature; the other five stages depend only on temperature. Soybeans flower sooner with long nights than with short ones. Some cultivars are more sensitive to night length than others, a feature which permits their adaptation to different latitudes. A phenological model was developed to permit the prediction of the timing and duration of various stages of development under a range of latitudes and planting dates. A response to GDD was used that was sensitive to the decreased rates of development at higher temperatures. Considerable care was given to the water relations of the soil–root system. Evapotranspiration was a function of solar radiation and temperature. Under the model, leaf and stem growth rates were reduced when the water supply to roots was less than 1.5 times the potential evapotranspiration rate. As water stress occurred there was a reduction of leaf and stem expansion, increased allocation of photosynthates to the roots, and a reduction of photosynthesis and transpiration.

This model, called SICM (for Soybean Integrated Crop Management), was developed by Jones et al. (1986) to study the effects of management practices and crop–pest interactions on soybean yield and profit. Some features of SICM are similar to SOYGRO, described earlier, but in other respects it is quite different since it is a management model. Submodels within SICM included the crop response to climate, soil, and water; to economics; and to several soybean pests. A variation called STRATEGY made use of historical climate data and expected market information to deveop management strategies.

The corn earworm submodel   The corn earworm submodel used by Stinner et al. is a population dynamics model divided into six developmental stages: eggs, small larvae, medium larvae, large larvae, pupae, and adults. The age structure was maintained within each developmental stage. The effect of temperature was determined by calculating "physiological days." A physiological day was defined as the proportion of development completed in one day at a reference temperature, taken to be 27°C. For example, at 18°C a corn earworm takes twice as long to develop from egg to adult as it does at 27°C; thus, one day at 18°C equals 0.5 physiological days. Variability of individual developmental rates within the poulation was accounted for in the model. The total number of eggs laid by a female corn earworm in her lifetime is a function of temperature, and the proportion of those eggs laid on any particular day is a function of temperature and adult stage. In the model, the number of eggs each female laid each day was calculated according to that day's temperature. The eggs were distributed on the soybean plants according to field data.

Mortality inflicted by predators, insecticides, and food shortages was modeled. For all stages except pupae, mortality was taken as a constant proportion per day throughout the season: 0.10, 0.05, 0.05, 0.02, and 0.05 for eggs, for small, medium, and large larvae, and for adults, respectively. Mortality for pupae was a function of crop leaf area index. Mortality due to food shortage was zero unless the leaf area index had been zero for at least two days, then increased linearly until it reached 1.0 ten days later if the leaf area index remained at zero.

For the model experiment, Stinner and colleagues selected Castana, Iowa, a site for which there is excellent climate data and soil information. Castana is well north of the area where the corn earworm is a major pest on soybeans, but it is within a major grain-producing region of the Midwest. Three sets of model simulations were done, using the actual climate data for Castana in 1979, the GISS $2 \times CO_2$ projections, and the GFDL $2 \times CO_2$ projections. In these simulations, a cultivar known as "Wayne" was used. It is an ideal choice for central Iowa today, but may not be with a warmer and drier future. The higher summer temperatures and diminished precipitation projected under the GISS and GFDL scenarios reduced the yield of the "Wayne" soybean cultivar and had a synergistic effect on the soybean–corn earworm interaction. The higher temperatures generated more rapid insect growth early in the season. This resulted in an increased, earlier emigration from corn to soybeans and increased damage. The GFDL scenario projected a much drier summer season and a much greater economic loss than did the GISS scenario. Stinner and colleagues suggested that many pests associated with soybeans will increase with the anticipated climate change and that increased use of pesticides will be essential to avoid economic disaster. They remind us that during the 1988 hot summer outbreak of the spider mite on soybeans, insecticide use increased tenfold.

Livestock and Climate Change

Animal products—meat, milk, and eggs—make up a significant part of the human diet. Fifty-five percent of farm products sold in the United States are from animals. Cattle production is greatest west of the Mississippi River and east of the Rocky Mountains. Six states, Texas, Iowa, Nebraska, Kansas, Oklahoma, and Missouri (in that order), contribute the highest number of cattle. The greatest swine production is in the Midwest, with Iowa being the leading state. Poultry are generally raised in confinement and near population centers. California is the leading producer of poultry.

Temperature, precipitation, humidity, and day length are important climate factors that have both direct and indirect effects on animal production. Climate has direct effects on animal metabolism, nutrient uptake, growth, reproduction, health, and general performance. It has indirect effects via crop and forage production for animal food, parasites, and animal diseases. Cattle, swine, and poultry are all susceptible to heat stress and drought (see Decker, 1983, and Decker et al., 1986). Low fertility rates, high mortality among newborns, weight loss in market-ready cattle and hogs, and other consequences may result from heat stress. Reduced egg and milk production is commonplace with stress. It has been suggested that maximum temperatures during summer in temperate regions approach survival limits for livestock.

Many animal diseases are kept in check largely as a result of climate restrictions on their vectors (intermediate organisms that carry the pathogen to the host), on their environment, or on the disease-causing agents themselves. Vector-borne diseases, zoonotic diseases (diseases that humans can catch from other animals), and animal diseases from other countries will be of increasing importance in the United States as the climate changes. Vector-borne diseases such as anaplasmosis, bluetongue, babesiosis, and Lyme disease are likely to increase with a warming climate. Zoonotic diseases such as Eastern equine encephalitis, intestinal nematodes (canine hookworm and roundworms), Q fever, toxoplasmosis, and aflatoxicosis will also increase as warmer temperatures lead to longer warm season. Poultry diseases such as salmonellosis, coccidiosis, influenza, and Newcastle disease will also increase in a warmer world.

As a part of the U.S. Environmental Protection Agency-supported studies, Stem et al. (1989) reported their analyses of the response of many of these diseases to the GISS and GFDL $2 \times CO_2$ climate scenarios. Many foreign animal diseases are likely to increase in the future as people and animals move around more. Wind can transport a disease such as foot-and-mouth disease and infect animals very quickly. Vector-borne foreign animal diseases such as African swine fever and Rift Valley fever are likely to increase in the United States as the habitat becomes more suitable for their vectors.

Vector-Borne Diseases

African swine fever   African swine fever is a virus carried by ticks. The particular tick involved could potentially occupy all of the United States south of a line from Oregon to Nebraska to Pennsylania under the projected $2 \times CO_2$ climate change. Today the same vector only occupies the southwestern U.S. from California to Texas.

Rift Valley fever   Rift Valley fever is a vector-borne viral disease of sheep and cattle, but it also affects humans, camels, monkeys, mice, rats, ferrets, hamsters, and goats. As many as 26 arthropods may act as vectors for Rift Valley fever within Africa. Several species of mosquitoes that are among its vectors in Africa are also found within the United States. One such mosquito, *Culex pipiens*, is distributed worldwide within a belt between 60° north latitude and 40° south latitude. As with most mosquitoes, it thrives in moist habitats. Development throughout its life span is best at temperatures between 25°C and 30°C; extended cold periods of 3°C or less is detrimental to its development. With global warming it is expected that many mosquito species will expand their distributions and therefore act as active vectors for Rift Valley fever and other diseases.

The Horn Fly

The horn fly is a ubiquitous pest of pastured cattle throughout the United States. These flies lay their eggs in cattle dung, and the adult horn flies feed on the blood of cattle. The cattle resort to head throwing, skin rippling, and tail switching in an attempt to fend off the biting flies. All this activity causes a reduction in weight gain and milk production, as well as less efficient feed conversion. The phenology and density of the horn fly is largely controlled by temperature. Adult horn flies begin their activity in the spring as temperatures warm, then produce many succeeding generations until cooler temperatures in the autumn trigger the overwintering diapause state.

Schmidtmann and Miller (1989) used a deterministic, climate-driven population model designed by Miller (1986) to estimate the effect of the GFDL $2 \times CO_2$ climate scenario on populations of the horn fly and its impact on the weight of beef cattle and the milk production of dairy cattle. The model was based on life-stage-specific growth and survival characteristics of the horn fly as a function of temperature and rainfall. They found that with climate warming there will be both increases and decreases of horn fly populations within the United States, depending on the region. Conditions favoring increases will occur in all areas except the southern states from Mississippi west to central California, where the fly will be suppressed during very warm summer months.

Throughout most of the United States the horn fly will emerge earlier in spring than it does now and will increase rapidly in numbers, then level off at densities higher than the current levels during the warmer greenhouse summers. Ambient temperature is the primary factor generating the projected increases in horn fly populations. The exposure of beef cattle to greater numbers of horn flies for longer periods will suppress weight gain, causing serious economic losses. The greatest losses will occur in the mid-Atlantic states, but the Northwest, Southeast, and Midwest regions will also incur large losses. Dairy cattle and milk production in the upper Midwest, Northeast, and mid-Atlantic regions would also be adversely affected by increased populations of horn flies with the warmer climate.

When ambient temperatures are above 34°C for more than two hours, generating cattle dung temperatures above 44°C in the sun, the immature horn flies are killed. Thus it was found using the GFDL $2\times CO_2$ scenario that the model projects a depression of horn fly density in the southern states west of the Mississippi. The spring and autumn climate conditions were found to be favorable to the horn fly in that area, but the summers generally would be too warm.

Drought

Water is essential to all life. The lack of water can greatly inhibit primary production, growth, productivity and, in fact, the success of all organisms. Water evaporates from one place and falls as precipitation in another. This movement of water in the atmosphere transfers energy from one part of the world to another and is a dominant determinant of the character of weather and climate. Too much water and the land erodes. Too little water and everything dries up.

Drought is the lack of sufficient water to meet basic needs. From a climatic standpoint, drought is a deficiency of moisture relative to a long-term average condition resulting from an imbalance between precipitation and potential evapotranspiration. Various types of drought are recognized, such as meteorological drought (a departure of precipitation from normal), agricultural drought (insufficient soil moisture for adequate crop growth), or hydrological drought (based on parameters such as stream flow). Drought is an event that may be characterized by its duration, intensity, and areal extent.

Drought has long been of interest to ecologists, but the effect of drought on crop production has no doubt received the greatest amount of attention. Drought is a major cause of yearly variability in crop production throughout the world. During the drought of the 1930s in the United States, yields of wheat and corn in the Great Plains dropped to as much as 50% below normal. During the 1950s drought the declines were less dramatic. During the drought of 1988, national corn yields were 40% below the average; catastrophic forest fires broke out in Yellowstone National Park, covering nearly 60% of its land area, and the water level of the lower Mississippi River was so low during June and July that barge traffic was stopped. (There was even a suggestion made at that time to divert waters from Lake Michigan into the Illinois and Mississippi Rivers.)

While the drought of 1988 afflicted about 40% of the United States, it was not as long-lasting as the droughts of the 1930s and 1950s. Droughts are often very spotty in areal extent. Usually drought will affect some areas of the world at any given time. In 1987, Pakistan and

India received less than half their normal rainfall as the summer monsoon weakened. Italy had a severe drought in 1988, and eastern China had its driest period in 114 years at that time. Drought in the western regions of the African Sahel continued from 1950 through the mid-1980s. A warmer world resulting from an increase of greenhouse gases will have many instances of reduced precipitation and the likelihood of longer and more intense droughts. However, no one is suggesting that any of the droughts experienced up to the present time are the direct result of greenhouse warming.

Droughts and the Solar Cycle

For many years, scientists have been trying to establish connections between solar activity and the Earth's weather and climate. Their attempts to do so have been largely frustrated by their inability to establish a firm physical linkage. However, this shortcoming has not stood in the way of attempts to correlate solar cycles with repetitive climate events. Global maps indicating possible rainfall variations with the solar cycle were drawn by Clayton (1923) and later by Shaw (1928). Their findings indicated greater rainfall near the equator and at high latitudes at times of high sunspot number; more recent studies have failed to find an 11-year cycle (see King, 1973).

A stronger correlation has been found between the Hale double sunspot cycle of about 22 years, the magnetic cycle of the sun, and periods of drought in the High Plains region of the western United States. Many investigators have suggested such a correlation, including Douglass (1919, 1928, 1936) and Abbot (1956). Mitchell et al. (1979) reconstructed the occurrence of droughts from tree-ring records for 40 geographic regions in the western states from 1600 to 1962 and found a quasi-periodicity of about 20–22 years. The maximum extent of the drought falls early in the double cycle, about 2 years after alternate minima. Currie (1981) has shown that for part of this period the recurrence of drought is closer to the 18.6-year period of revolution of the nodes of the lunar orbit. An excellent review of solar variability, weather, and climate is found in a report published by the National Research Council, Geophysics Study Committee (1982).

Droughts, Pressure Systems, and El Niño

The summer of 1988 was a hot and very dry one throughout the north-central region, the west coast, and the southeastern United States. In October 1986, soil moisture content was above average over most of the United States, although not in the Southeast. By April 1987, the dry conditions in the Southeast had moderated, but the development of drought on the West Coast was well under way. By April 1988, drought conditions were pervasive along the West Coast and northern U.S. and were returning to the Southeast. 1987 was a time of an El Niño event, with sea surface temperatures above normal along the equator from South America to the central Pacific Ocean. Associated with this condition are large displacements of the major rain-producing convergence zones in the tropics and atmospheric circulation changes. During northern winter months, such conditions favor the development of a stronger than normal ridge of high pressure near the west coast of the U.S. and lower pressures in the Aleutian Low over the north Pacific Ocean. In the aftermath of El Niño, exceptionally cold water arrived in the tropical Pacific Ocean, an event known as La Niña. El Niño and La Niña are discussed in more detail later in this chapter.

Studies of many great U.S. droughts of this century have demonstrated that summer drought is a possible consequence of three upper-level high-pressure cells forming over the North Pacific Ocean, the North Atlantic Ocean, and the United States, with contiguous low-pressure cells between them. Trenberth et al. (1988) showed, with the use of numerical models, that the three-cell flow pattern in the upper atmosphere was related to the establishment of abnormally cold water in the tropical North Pacific Ocean. As Trenberth and colleagues clearly say, and is pointed out by Namias (1989), this is not the sole cause of the drought and other subtle events may be related. According to Namias, many major North American droughts of this century have not been associated with La Niña-type conditions, although the three-celled pattern has occurred. A drying out of soil in the spring in the Midwest assists in generating a high-pressure cell and promotes the drought. Normally during summer months, the jet stream flows across the United States near the Canadian border and is quite flat, exhibiting relatively little amplitude. But when these distinctive high- and low-pressure cells become established, they create large undulations in the jet stream. The result is the occurrence of droughts under the high-pressure ridges and heavier-than-normal rainfall in the low-pressure troughs.

Desertification

A drought may not lead to desertification, but if a dry spell in a given area is persistent and extreme, and if overgrazing by cattle or human activities is superimposed, then whatever vegetation was growing there may be diminished beyond recovery. Charney et al. (1975) first proposed a biogeophysical feedback mechanism that tends to produce changes in rainfall and plant cover. Ground covered by plants has an albedo in the range 10% to 25%, whereas dry, light, sandy soil may have an albedo as high as 35% to 45%. In such instances, a decrease of plant cover would result in an increase of albedo, a decrease in the net incoming radiation, and an increase in radiative cooling of the air. The cooler air would begin to sink. This in turn would suppress cumulus convection over the region and there would be reduced rainfall. Lower rainfall would have an adverse effect on plants and lead to a further decrease of plant cover. This positive feedback would promote desertification. Charney and colleagues used the GISS GCM to apply these ideas to a calculation of the climate of the Sahara, and in particular of the southern Sahara, the Sahel. They concluded that this mechanism is a plausible one for promoting desertification.

In addition to the desertification mechanism proposed by Charney, according to Schlesinger et al. (1990) there are additional ecosystem feedbacks that lead to desertification once the process is started. Schlesinger and colleagues studied ecosystem processes on the Jornada Experimental Range in southern New Mexico and found that long-term grazing on semiarid grasslands leads to an increase in the spatial and temporal heterogeneity of water, nitrogen, and other soil resources. During the last 100 years, large areas of black grama grass have been replaced by shrub communities of creosote bush and mesquite. Black grama grass is shallow rooted and forms an effective shield against soil erosion by rainfall. In these semiarid grasslands, most of the moisture and biotic processes are constrained to the upper soil layers, including mineralization and nitrogen cycling. Large herds of domestic livestock disrupt this tightly knit surface ecosystem and lead to a decline of black grama grass. Trampling results in the compaction of soil and reduces infiltration of rainfall. Greater runoff leads to soil erosion and the transport of water, nitrogen, and other nutrients into low areas and streambeds. Shrubs invade those areas since they can more easily exploit the additional soil moisture that does infiltrate in some places. Shrub dominanace is heterogenous and the more barren intershrub spaces allow soil erosion by wind and water, and nutrient loss.

When shrubs replace grassland, a greater percentage of the soil is bare, and the temperatures of the soil surface and the air near the ground increase, even though the albedo of the soil surface is greater than it was with the grass cover. Hot, dry soils retard the accumulation of organic nitrogen in the soil and this can lead to more sparse shrub growth, since shrubs are less dependent on the surface nitrogen supply. As these abiotic processes dominate over biotic ones, the ecosystem tends toward aridity. Once begun, this process of heterogeneity in arid regions is likely to develop a positive feedback and further exacerbate the trend toward desertification. Arid lands now cover about 12% of the Earth's land surface, but when semiarid grasslands and woodlands are included, the total extent of dryland ecosystems is about 33%.

Satellite images show that arid lands in southern New Mexico and the Sahel are expanding. Greenhouse warming may extend these dryland areas greatly and promote them in many other parts of the world.

When desertification occurs, less solar radiation is used in evapotranspiration, and daytime surface air temperatures increase. Balling (1991) showed that widespread desertification around the world, notably the Sonoran, Sahel, Turkestan, and Gobi arid regions, is producing an additional statistically significant warming trend in land-based temperature records over the past century. Areas undergoing desertification have an average warming trend of 0.05°C per decade when compared to nearby areas with no desertification. This suggests further that any greenhouse-driven desertification may amplify regional, and maybe even global, warming.

El Niño and La Niña

Floods in Ecuador and Peru, failure of the Peruvian anchovy harvest, crop failures and drought in eastern Australia, a mosquito-borne disease outbreak in Australia, drought in southern India and southern Africa, and even vicious storms in California and widespread flooding in the southern United States, may repeatedly have been consequences of a complex synchrony between two closely coupled events—one in the Pacific Ocean, known as El Niño, and the other in the atmosphere, known as the Southern Oscillation. Together they are referred to as ENSO and their effects are truly global. They reoccur with a periodicity of from 2 to 7 years.

The coast of Peru has one of the driest deserts in the world, the Atacama. The adjacent cold waters of the Pacific Ocean are normally extremely productive of fish, especially anchovies, and support a huge bird population. A warm southward-flowing current often occurs off Ecuador and Peru around Christmastime. It was named El Niño by the Spanish-speaking people living there, meaning "boy child," for the Christ child Jesus.

Every few years unusually warm water appears off South America, extending westward to extraordinary distances along the equator, often as far as the international date line and beyond. Although this warm water is not really related to the annual warm current, it has assumed its name, El Niño. Each time this great ocean warming occurs, the barometric pressure over vast reaches of the southeastern Pacific falls while the pressure over Indonesia and northern Australia rises. When an El Niño ends and cold water again appears off South America, an event referred to as La Niña, meaning "girl child," the pressure difference reverses, with the pressure becoming high over the eastern Pacific and low over the western Pacific and Indian oceans. Sir Gilbert Walker,

Director General of Observatories in India, trying to explain the failure of the monsoons in 1904, became aware of the pressure differences occurring in these regions and the fact that they reversed every few years. Sir Gilbert named this seesaw or flip-flop the Southern Oscillation. Climatologists use the atmospheric pressure difference between Tahiti and Darwin (in northern Australia) as an index of this seesaw of the ocean and atmosphere, the ENSO.

One would expect the atmosphere to respond to a pressure difference much more quickly than the massive ocean. Water normally sloshes from the western Pacific to the eastern Pacific and back again. However, the water is driven by surface winds that respond to gradients of atmospheric pressure. A meteorologist, Jacob Bjerknes (1969), first demonstrated a link between the Southern Oscillation and sea surface temperatures in the Pacific Ocean and related the ENSO to weather events in the North Pacific and North America. Bjerknes coined the term "teleconnection" to describe the tight coupling of ocean and atmospheric dynamics, and demonstrated clearly that the ocean dynamics acted as a "flywheel" on the atmosphere. He showed that an initial change in the ocean could affect the atmosphere in a manner that would in turn induce further changes in ocean circulation, reinforcing its initial trend. For example, a slight relaxation of the easterly trade winds (flowing from the east), which drive warm surface waters westward and bring up cold water along the Peruvian coast, would reverse that trend and cause a modest warming of the central and eastern Pacific. This warming would then cause a further relaxation of the trade winds, and further warming, so that an El Niño develops. This theory of the ENSO has been considerably refined during recent years; excellent reviews are given by Rasmussen (1985) and by Philander (1989).

Normally the trade winds over the northeast and southeast Pacific Ocean converge just north of the equator, along a line called the intertropical convergence zone. These trade winds flow westward near the equator into a semi-permanent low-pressure region in the vicinity of Indonesia. When the Southern Oscillation index (Tahiti pressure minus Darwin pressure) is high, the east–west pressure gradient along the equator increases and the westward flow of air strengthens, dragging the surface water westward. As a result, sea level rises in the west and lowers in the east. The piling up of warmer water in the western Pacific forces deep, colder water to rise near the surface off Ecuador and Peru; this water, rich in nutrients, supports the abundant fish and bird populations. When the Southern Oscillation index falls, pressure rises over Indonesia and northern Australia, and the pressure gradient along the equator decreases or even reverses. The normal equatorial easterly trade winds become westerly, dragging the surface ocean waters eastward and pushing warm water toward South America. The ocean resists in the form of internal waves, known as Kelvin waves, which take many weeks to cross the Pacific, but as they reach the South American coast they depress the colder water and replace La Niña with El Niño. The lack of nutrients in the warm surface waters then stifles the ecosystem productivity, the fisheries fail, and the seabirds starve.

During the winter of 1982–1983, the sea surface temperature rose as much as 14°C above normal as the largest modern El Niño on record developed. The Peruvian fishery supplies more than one-third of the world's fish meal, an important fertilizer and animal feed. Not only was that year a disaster for the fishery off South America, but there was disruption of marine ecosystems from the Gulf of Alaska to southern Chile. Barracuda appeared off the Oregon coast, bluefin tuna were found off British Columbia, and salmon moved to cooler water farther north. Some marine life shifted northward by as much as 1000 kilometers with the warming sea surface temperatures. Winter storms off the southern California coast destroyed much of the giant kelp canopy. The entire seabird population of Christmas Island disappeared because of the higher sea level and the destruction of fish stocks and squid upon which the birds depended (see Schreiber and Schreiber, 1984). Coral mortality was high along the eastern equatorial Pacific coast. Torrential rains in Ecuador and Peru produced widespread damage, as did flooding in Lousiana, Florida, and Cuba. Snowpack was at record levels in the Sierra Nevada and Wasatch Mountains, producing flooding during the spring runoff. At the same time, droughts occurred in Indonesia, the Philippines, southern India, and southern Australia.

The teleconnections are clearly in evidence around the world, although every El Niño does not produce the same weather events. There is almost always heavy rain in Peru and usually dryness in Australia, but the effects in other places are not often as consistent. Handler and Handler (1983) studied the relationship of United States corn yields with 49 El Niño events occurring over a 112-year period. Because some of the events came too late to affect yields that same year, they ended up using 31 warming events. Years with a significant El Niño were years with higher than normal corn yields. The 1991 El Niño moved the subtropical jet stream northward across the southern United States and torrential rains soaked southern Texas, causing record floods and enormous damage.

Searching for a clue as to why intervals between strong El Niño events range from 2 to 7 years and average 4 to 5 years, Rasmussen et al. (1990) have found El Niño to be paced by two different clocks in the Pacific equatorial winds, one with a 2-year beat and the other with a less well defined beat of 4 to 5 years. There is an interplay between the biennial cycle and the weaker, longer period cycle. When they are in phase, a strong El Niño may result, and when they are out of phase the El Niño will be weak. Just why these periodicities occur is not entirely understood and theoreticians have yet to agree on how to tie together the atmospheric and oceanic waves into a complete model.

CHAPTER EIGHT

What to Do?

THE WORLD IS CHANGING as a burgeoning human popu-
lation overuses available resources, depletes soils, and contaminates
water, land, and atmosphere. There is no place on Earth untouched by
human activity as the global ecosystem slowly but inexorably degrades.
Scarcely an ecosystem on Earth resembles its original state prior to
human encroachment. Amazingly, the world has seemed well buffered
against some of these impacts—yet this is the insidious part of global
change. It proceeds slowly, subtly, and yet is constantly ongoing toward
a less productive state.

Life on Earth has always been subjected to change, but the changes
humans are imposing on global ecosystems are immediate, substantial,
and sustained. Violence has always been a part of the natural world.
Windstorms, hurricanes, tornados, fires, floods, and droughts are epi-
sodic, chaotic, and patchy. But human disturbance is continuous, or-
ganized, and broadscale.

With greenhouse warming, natural ecosystems will sustain vast
changes as older trees die off, some seedlings fail to mature, pathogens
and predators succeed, some reproduction fails, and competition among
species shifts. Warmer-adapted species may be able to replace colder-
adapted species, but in many situations they may not be able to reach
a favorable habitat. Nearly every analysis presented in the foregoing
chapters projected a biomass loss from global warming, even with CO_2
enrichment facilitating plant growth. In some respects, natural unmanaged ecosystems are more vulnerable to climate change than are managed ecosystems. Agricultural management will include the use of new
genetic hybrids selected for changing conditions, and possibly improved
irrigation and nutrient enrichment practices. However, it would be folly
to assume that technology will solve the agricultural problems created
by global warming, and the economic cost of maintaining adequate crop
productivity could be immense. The greatest climate and ecosystem

246

changes are likely to occur at high latitudes, where the projected boreal forest displacement is estimated to be substantial. Mean annual global warming may average 3°C with a $2 \times CO_2$ climate change, but high latitudes may increase as much as 6°C to 8°C, all in less than 70 years. But that is not the end of it, since greenhouse gas emissions will continue to accumulate in the atmosphere for a long time beyond the equivalent carbon dioxide doubling level.

Many of the effects of global warming are unknown, particularly those relating to water availability. The world as a whole may become wetter as it warms, and that could be for the better. However, increased wetness in some regions may produce breeding areas for more human and animal disease vectors. Wetness also suggests more bacteria, fungi, and other pathogens. Increased respiration by most organisms may result from both warming and wetness, thereby contributing a positive feedback to the greenhouse gas-driven climate change. Some regions are expected to become drier than they are today, including some of the most agriculturally productive areas of the world. Warmer and longer growing seasons mean increased evapotranspiration by all plants and a greater demand for water. This change will be somewhat offset by increased water use efficiency by plants benefiting from enriched atmospheric carbon dioxide. A warmer world will take a toll in human health; particularly vulnerable are the chronically ill, the elderly, and the very young. Combined stresses of heat and pollution will be especially deleterious to humans.

The increase of the annual global air temperature of 3°C above the present mean, projected under the equivalent of carbon dioxide doubling, will make the world warmer than it has been at any time in the last 10,000 years. In fact, one must go back to the Emian interglacial period 130,000 years ago to find the Earth 3°C warmer than it is today. It would be serious enough to have only carbon dioxide increasing in the atmosphere as the result of various human activities, but to have an additional five molecular species contributing to global warming is alarming, to say the least. (The greenhouse gases and their relative effectiveness, concentrations, annual rates of increase, and current contributions to climate warming were listed in Table 1 of Chapter 1.) Predicting the increase in greenhouse gases for the future is difficult. Emission rates from all sources are uncertain. Will rising temperatures increase the rate of release of carbon dioxide by respiration, or suddenly release methane that has been buried in polar regions? We do not know. Nor do we understand the role the oceans may play as a sink for carbon dioxide. Nevertheless, all measures show an upward trend in the concentration of all greenhouse gases in the atmosphere. A best-estimate, business-as-usual scenario reported by Leggett (1990) shows the equivalent carbon dioxide concentration—the total effectiveness of all greenhouse gases taken together—will have doubled from its preindustrial level by the year 2030. At the present time, the equivalent carbon dioxide concentration is rising at a rate of 6% to 8% per decade. Since greenhouse gases accumulate in the atmosphere, the growth is compounded like interest in a bank account. However, the accumulated effectiveness of the gases grows more rapidly than just compounded interest since all the other gas species are more effective than carbon dioxide.

Strategies

It is never too late and it is always too late to begin mitigation of the greenhouse effect. The earlier action is taken to reduce the emission rates of greenhouse gases, the less drastic it will need to be and the more effective will be its consequences in the long term.

Kelly (1990) discusses several possible scenarios for reducing the rate of greenhouse warming. The first steps toward reducing greenhouse gas emissions would be those that meet two basic criteria: measures that also help to alleviate other environmental problems, and measures that result in minimal societal impacts. Phasing out the production of chlorofluorocarbons (CFCs) by 1997, in agreement with the Montreal Protocol of 1987, and a halt to deforestation by the year 2000 would be the first such necessary steps. These actions would delay the date of equivalent carbon dioxide concentration doubling by ten years.

What would a 50% reduction in greenhouse emissions accomplish? In order to accomplish this, the following actions are needed: Eliminate chlorofluorocarbon production by 1997; halt deforestation by the year 2000; and reduce fossil fuel carbon emissions by 20% by the year 2005 and by 50% by the year 2030. These actions would delay until late in the twenty-first century the time when the equivalent carbon dioxide concentration would double from its preindustrial value and would reduce the rate of warming by more than one-third. However, even with this scenario, the equivalent carbon dioxide level would continue to rise.

Finally, what would it take to stabilize the equivalent carbon dioxide level in the atmosphere at a predetermined level, say, at 375 ppm? A possible stabilizing scenario would be as follows: Eliminate chlorofluorocarbon production by the year 1997; halt deforestation by the year 2000, followed by massive reforestation, enough to absorb 1.65 Gt (gigatons) of carbon by the year 2020; reduce fossil fuel carbon emissions by 70% by the year 2020; and reduce the annual rise of methane and nitrous oxide concentrations by 75% by the year 2020. These measures are drastic, but would stabilize the equivalent concentration of carbon dioxide at 375 ppm during the 2020s. With this scenario, temperatures are projected to rise only 2°C or less by late in the twenty-first century.

These are a few examples of many potential strategies for reducing greenhouse gas emissions. In any event, these actions are only a beginning, and other actions may be needed in the long run. Concern over the destruction of stratospheric ozone and the resulting increase in ultraviolet solar radiation has motivated national governments to agree on a phaseout of all chlorofluorocarbons before the twenty-first century. Here is an example of immediate action being agreed to as a consquence of fear of impending disaster to life. But the agreement to phase out CFCs was a simple decision compared with achieving a consensus to reduce the use of fossil fuels in a world while massive amounts of coal remain in the ground. Among all of these actions, reforestation of large regions is very attractive. However, the numbers involved in doing this are massive. To achieve the goal of sequestering 1.65 Gt of carbon per year, 10 million hectares of forest, an area almost the size of New York State, would need to be planted each year for the first two decades of the twenty-first century. Other estimates are similar in terms of the massive amount of reforestation necessary to absorb a significant amount of carbon dioxide from the atmosphere. Furthermore, the assimilation process would only be useful for a few decades; after that the forest would be cut and the wood stored in some nonrespiring manner. Certainly planting vast numbers of trees is highly desirable for many reasons, but the likelihood of this process making a very significant dent in the rate of increase of the atmospheric carbon dioxide concentration is not very good.

Technological Fixes

Various technological schemes have been suggested for mitigating global warming. One idea is to place huge mirrors in space to reflect sunlight. To achieve a 3.5% reduction of incoming sunlight, equivalent to compensating for a temperature increase of 2.5°C, a minimum mirror area of 4.5×10^6 km^2 would be required, and it would have to be positioned in space so as to always shade the Earth from the sun. Not very feasible.

Another scheme for increasing the planetary albedo and thereby reflecting more sunlight to space is to increase the reflectivity of marine stratocumulus clouds by about 4%; this would approximately compensate for the greenhouse warming from a doubling of atmospheric carbon dioxide. To achieve this, it is estimated that naturally occurring cloud condensation nuclei (CCN), upon which cloud droplets form, would need to be increased by about 30%. Most of these CCN are sulfuric acid particles formed by the oxidation of atmospheric sulfur. Adding 30% more sulfuric acid droplets to the atmosphere would require 32,000 metric tons *per day* of sulfur dioxide. To minimize possible deleterious effects, including acid deposition, the sulfur dioxide would need to be released over the oceans, downwind of land masses. This scheme requires the location of many large coal-burning power plants along coastlines. The coal they burn would be especially enriched with sulfur. Not a very practical idea.

Some have dubbed the next scheme the "Geritol" solution for global warming. The oceans have been mentioned as playing a key role in the carbon cycle, but it is a role not well understood. The Southern Ocean around Antarctica and parts of the equatorial Pacific are rich in nutrients, but poor in iron and low in phytoplankton productivity. When small amounts of iron were added to samples of water containing phytoplankton, productivity soared. Phytoplankton productivity is part of the biological pump that removes carbon dioxide from the atmosphere and sequesters it in the deep ocean. If the Southern Ocean could be fertilized with iron, then the absorption of atmospheric carbon dioxide might be greatly enhanced. Joos et al. (1991) calculate that if iron were continually added for 100 years to the phosphate-rich waters of the Southern Ocean, corresponding to 16% of the global ocean surface, the atmospheric CO_2 concentration would be 59 ppm below what it would have been without fertilization with no increase in anthropogenic CO_2 emissions. If the CO_2 emissions are included in the calculation, then the phytoplankton uptake would reduce atmospheric CO_2 concentration by 90–107 ppm. Proponents admit that such a large uptake of CO_2 is unlikely to be achieved in practice. Iron enrichment could greatly alter the composition of phytoplankton species in the Southern Ocean and change the entire food chain. Increased productivity might render the deep ocean oxygen-deficient and lead to an increase in nitrous oxide and methane gases released to the atmosphere. Although the amounts of iron required are not stupendous, the idea does not appear very feasible until much more is learned.

Many technological mitigating methods have been proposed for minimizing the emission of carbon dioxide into the atmosphere. Most of these methods attack the problem at the power plant stack, using physical and chemical reduction techniques. They involve the absorption of flue gas carbon dioxide by liquid or solid solvents, the possible reduction of CO_2 to a synthetic fuel using a non-fossil fuel source, extraction of CO_2 with lime during combustion, and other chemical processes. All are extremely energy intensive and impractical in the foreseeable future.

Into the Future

The world is warming and will continue to warm far into the future. We know that to slow global warming and avert its many attendant ramifications, we need to reduce the emission of greenhouse gases. Clearly, one way to accomplish this is to reduce the use of fossil fuels. This can be done by using alternative energy sources such as solar, wind, biomass, ocean thermal electric conversion, geothermal, tidal, and wave energy. Another obvious choice is to develop passively safe nuclear reactors. But the nuclear option is a difficult one for society. Concerns about nuclear power plant safety and the disposal of radioactive waste dominate the nuclear debate. The nuclear option does not seem viable until these concerns are laid to rest.

There is another option—in many respects it is the most obvious, urgent, and important of all. That option is conservation of energy. This is an option society can take immediately, and by taking it buy time against rapid climate change. The technology for implementing vast energy efficiency measures is known today; we need not wait for future technological developments to take place. The energy conservation and efficiency option is economically sensible, resource wise, and environmentally comfortable.

Any plan for the future dealing with global warming must include ideas for coping with this slowly evolving condition. How to maintain agricultural productivity? How to manage forests in a continuing state of health? How to adjust for rising sea levels and coastal inundation? How to help all peoples of the world understand the problem and how they themselves can manage to live with climate change and, at the same time, understand what to do about it.

A majority of, if not all, nations of the world will need to agree on policy actions for moderating greenhouse warming. The achievement of such action is complicated by political, social, economic, and demographic differences among nations. These complexities suggest that action may develop slowly. However, the problem of global warming is a continuing one, and one which will not fade from the consciousness of humankind. Eventually, all people must realize that the world can sustain a healthy and viable condition for a very long time into the future provided they work together.

Glossary

Aerosols Small particles, usually droplets, suspended in the atmosphere.

Air mass An entity of the atmosphere that is approximately homogenous in its horizontal distribution of temperature, humidity, and lapse rate.

Albedo The fraction of incident light reflected by a surface. It refers to the "whiteness" of the surface. Originally it meant only the reflection of visible wavelengths of light, but it is used more generally to mean the reflection of all wavelengths.

Anthropogenic Resulting from human activities.

Bioclimatic systems A group of items involving climate and organisms.

Biomass The total weight of all the living organisms, or some designated group of the living organisms, in a given area.

Biome One of a number of distinctive ecological communities of the Earth, characterized primarily by the nature of its vegetation.

C_3, C_4 plants Refers to plant metabolic pathways for the assimilation of carbon, which involve either a 3-carbon or a 4-carbon acid. C_4 plants are usually adapted to warmer, drier environments than C_3 plants.

Cavitation The formation of air bubbles in liquid.

Clathrates Lattice-like structures in which a gas (such as methane) is trapped in a matrix of water molecules. Also known as a **hydrates**.

Conductance The readiness of water vapor to move out of and CO_2 to move into or out of a leaf.

Convection The transfer of heat within a fluid by mass movement that is generated by either a temperature or pressure difference.

Cyclone A storm or wind system that rotates counterclockwise (in the Northern Hemisphere) around a central area of low atmospheric pressure. In the Northern Hemisphere, the winds of an **anticyclone** rotate clockwise.

Downburst A powerful downward rush of air from an advancing storm.

Ecosystem   An ecological unit consisting of a particular habitat and all the organisms that live in it; a landscape unit that is relatively homogenous and reasonably distinct from adjacent areas.

Ecotone   A transition region between two distinct ecosystems.

Evapotranspiration   The loss of water from soil and and plants via evaporation and transpiration.

Equivalent CO_2 concentration   Producing a greenhouse effect equal to that caused by a given concentration of carbon dioxide.

Feedback   The return to the input of part of the output of a system or process. A **positive feedback** reinforces the output, whereas a **negative feedback** reduces the output.

Forcing factors   A quantity generating a change of something. As used here, it is a change imposed on the Earth's energy balance that alters the global temperature.

Forest gap model   A mathematical simulation of the processes of succession for a community of trees involving seed germination, establishment, growth, competition, and death.

Front   The boundary between air masses of different temperature, humidity, and lapse rates.

Gaia hypothesis   The idea that the biosphere is a self-regulating entity with the capacity to keep our planet healthy by controlling the chemical and physical environment. Named for the ancient Greek concept of "Mother Earth."

Greenhouse gases   Molecules made up of three or more atoms that absorb sunlight at infrared wavelengths; their presence in the Earth's atmosphere warms it.

Holocene   The most recent geological time period; generally the last 12,500 years. (Also known as the **Recent**.)

Instar   A stage in the metamorphosis of an insect.

Isogram   A line on a map or chart along which there is a constant value (of temperature, pressure, moisture, etc.). Also called an **isoline**.

Isotherm   A line of constant temperature on a map or chart.

Jet stream   A long, narrow, meandering current of high-speed winds near the tropopause. A jet stream generally blows from west to east.

Lapse rate   The rate of temperature change with increasing altitude.

Mesic   Refers to a moderate amount of water.

Monsoon   Land and sea winds of the Indian Ocean that change seasonally rather than daily; often characterized by heavy rainfall.

Mycorrhizae   Mutually beneficial associations of fungi with plant roots. Many plants, including almost all tree species, depend on mycorrhizae to increase their absorption of water and nutrients.

Paludification   A plant successional process leading to the formation of peatland.

Palynology   A branch of botany dealing with pollen and spores.

Phenology   The study of periodic biological events and their relationship to seasonal climate changes.

Phytoplankton   Minute forms of plant life in water.

Pleistocene   The geological time period that preceded the Holocene, lasting from roughly two million years ago to about 12,500 years ago.

Primary production   The transformation of energy by plants (and some microorganisms) from sunlight and inorganic chemicals into the chemical energy of organic compounds that serve as food for other organisms.

Sink   Something that acts as a storage point; a place into which energy or a substance may flow and remain.

Source   Point of origin of energy or a substance.

Stomata   The pores in a leaf through which water vapor and carbon dioxide gas may pass.

Thermocline   A layer in a thermally stratified lake that separates an upper, warmer, lighter, oxygen-rich zone from a lower, colder, heavier, oxygen-poor zone.

Water deficit   Demand for water that is not met by the available supply; a lack of water below a normal amount.

Xeric   Referring to a small amount of moisture; a very dry environment.

Acronyms and Abbreviations

2×CO₂ Doubled equivalent carbon dioxide concentration from a given reference level, such as the preindustrial level or the present-day level.

4×CO₂ Quadrupled equivalent carbon dioxide concentration from a given reference level.

APETR Annual potential evapotranspiration ratio.

APPT Average total annual precipitation.

Box's vegetation model parameters:
   DTY   Annual range of monthly average temperature.
   MI   Ratio of annual precipitation to potential evapotranspiration.
   PMIN   Minimum monthly average precipitation.
   PMAX   Maximum monthly average precipitation.
   PRCP   Annual precipitation.
   PMTMAX   Average precipitation of the warmest month.
   TMIN   Minimum monthly average temperature.
   TMAX   Maximum monthly average temperature.

CERES A growth and yield model for cereal grains.

CFC Chlorofluorocarbon.

COHMAP Cooperative Holocene Mapping Project.

DU Dobson units; the amount of ozone in a vertical column of the atmosphere measured in milliatmospheres per cm.

EPA Environmental Protection Agency; an agency of the United States government.

EPIC Erosion Productivity Impact Calculator; a model of primary crop production.

FORENA Forests of Eastern North America; a forest gap model.

FORET Forests of East Tennessee; a forest gap model.
   GAUSS   subroutine to randomly select GDD.
   KILL   subroutine for death of a tree.
   SPROUT   subroutine for vegetative reproduction.

FORSKA A boreal forest gap model.

GCM Global climate model or general circulation model.

GDD Growing degree-days.

GFDL Geophysical Fluid Dynamics Laboratory; source of GCM climate scenarios.

GISS Goddard Institute of Space Studies; source of GCM climate scenarios.

GVM Global Vegetation Model; a forest gap model.

   DYNAMIC subroutine at the taxonomic resolution of plant fundamental types.

   SPATIAL subroutine with a static grid containing restraints to growth.

JABOWA Botkin-Janak-Wallis model for simulating forest growth; a forest gap model.

   GROW subroutine for tree growth.

   BIRTH subroutine for seedlings to begin growth.

   SITE subroutine describing site characteristics.

kya Thousands of years ago.

MINK The region comprising Missouri, Iowa, Nebraska, and Kansas.

MIZ Marginal ice zone.

OSU Oregon State University; source of GCM climate scenarios.

ppb Parts per billion.

ppm Parts per million.

PSC Polar stratospheric clouds

SOYGRO Growth and yield model for soybeans.

TIMBER! A modified JABOWA model for forests of the Great Lakes region.

TOMS Total ozone mapping spectrometer.

USDA United States Department of Agriculture.

UV Ultraviolet radiation.

VPD Vapor pressure deficit.

W/m^2 Watts per square meter.

WUE Water use efficiency.

ya Years ago.

References

Abbot, C. B. 1956. Periods related to 273 months or 22 and 3/4 years. Smithsonian Misc. Coll. 134: 1.

Abrams, M. D. 1992. Fire and the development of oak forests. BioScience 42: 346–353.

Allen, R. G. and F. N. Gichuki. 1989. Effects of projected CO_2 induced climatic changes on irrigation water requirements in the Great Plains states (Texas, Oklahoma, Kansas, and Nebraska). In Smith and Tirpak 1989, Appendix C, pp. 6-1 to 6-42.

Anderson, J. G., D. W. Toohey and W. H. Brune. 1991. Free radicals within the Antarctic vortex: The role of CFCs in Antarctic ozone loss. Science 251: 39–46.

Andreae, M. O. 1986. The oceans as a source of biogenic gases. Oceanus 29: 27–35.

Andrewartha, H. G. 1961. Introduction to the Study of Animal Populations. University of Chicago Press, Chicago.

Andrewartha, H. G. and L. C. Birch. 1954. The Distribution and Abundance of Animals. University of Chicago Press, Chicago.

Angell, J. K. 1986. Annual and seasonal global temperature changes in the troposphere and low stratosphere, 1960–1985. U.S. Department of Agriculture Monthly Weather Review 114: 1922–1930.

Appenzeller, T. 1991. Ozone loss hits us where we live. Science 254: 645.

Arris, L. L. and P. S. Eagleson. 1989. Evidence of a physiological basis for the boreal–deciduous forest ecotone in North America. Vegetatio 82: 55–58.

Atkinson, R. J., W. A. Mathews, P. A. Newman and R. A. Plumb. 1989. Evidence of the mid-latitude impact of Antarctic ozone depletion. Nature 340: 290–293.

Atkinson, T. C., K. R. Briffa and G. R. Coope. 1987. Seasonal temperatures in Britain during the past 22,000 years reconstructed using beetle remains. Nature 325: 587–592.

Auclair, A. N. D. 1989. Climate theory of forest decline. In Proceedings of the IUFRO Conference on Woody Plant Growth in a Changing Physical and Chemical Environment, 27–31 July 1987, ed. D. P. Lavender. University of British Columbia, Vancouver. British Columbia, Vancouver, B.C.

Auclair, A. N. D., H. C. Martin and S. L. Walker. 1990. A case study of forest decline in western Canada and the adjacent United States. Water Air Soil Pollut. 53: 13–31.

Austin, M. P. and O. B. Williams. 1988. Influence of climate and community composition in the population demography of pasture species in semi-arid Australia. Vegetatio 77: 43–49.

Bakun, A. 1990. Global climate change and intensification of coastal ocean upwelling. Science 247: 198–201.

Ball, T. F. 1986. Historical evidence and climatic implications of a shift in the boreal forest–tundra transition in central Canada. Clim. Change 8: 123–134.

Balling, R. C. Jr. 1991. Impact of desertification on regional and global warming. Bull. Am. Meteorol. Soc. 72: 232–234.

Bartlein, P. J., I. C. Prentice and T. Webb. 1986. Climatic response surfaces from pollen data for some eastern North American taxa. J. Biogeog. 13: 35–57.

Batt, B. D., M. G. Anderson, C. D. Anderson and F. D. Casewell. 1989. The use of prairie potholes by North American Ducks. In Northern Prairie Wetlands, ed. A. G. van der Valk. Iowa State University Press, Ames. pp. 204–227.

Bazzaz, F. A. 1990. The response of natural ecosystems to the rising global CO_2 levels. Annu. Rev. Ecol. Syst. 21: 167–196.

Bernabo, J. C. and T. Webb. 1977. Changing patterns in the Holocene pollen record of northeastern North America: A mapped summary. Quat. Res. 8: 64–96.

Bjerknes, J. 1969. Atmospheric teleconnections from the equatorial Pacific. U.S. Department of Agriculture

Monthly Weather Review 97: 163–172.

Blasing, G. B. and A. M. Solomon. 1983. Response of the North American corn belt to climate warming. U.S. Department of Energy Report TR006. Washington, D.C.

Blumberg, A. F. and D. M. DiToro. 1990. Effects of climate warming on disolved oxygen concentrations in Lake Erie. Trans. Am. Fish. Soc. 119: 210–223.

Blumenthaler, M. and W. Ambach. 1990. Indication of increasing solar ultraviolet-B radiation flux in alpine regions. Science 248: 206–208.

Bonan, G. B. 1991. A biophysical solar energy budget analysis of soil temperature in the boreal forest of interior Alaska. Water Resources Res. 27: 767–781.

Bonan, G. B. and H. H. Shugart. 1989. Environmental factors and ecological processes in boreal forests. Annu. Rev. Ecol. System. 20: 1–28.

Bonan, G. B., H. H. Shugart and D. L. Urban. 1989. The sensitivity of some high-latitude boreal forests to climatic parameters. Clim. Change 16: 9–29.

Botkin, D. B. 1972. Some ecological consequences of a computer model of forest growth. J. Ecol. 60: 849-872.

Botkin, D. B., J. F. Janak and J. R. Wallis. 1970. A simulator for northeastern forest growth. Research Report 3140. IBM Thomas J. Watson Research Center, Yorktown Heights, NY.

Botkin, D. B., R. A. Nisbet and T. E. Reynolds. 1989. Effects of climate change on forests of the Great Lakes states. In Smith and Tirpak, 1989, Appendix D, pp. 2-1 to 2-31.

Botkin, D. B., D. A. Woodby and R. A. Nisbet. 1991. Kirtland's warbler habitats: A possible early indicator of climatic warming. Biol. Cons. 56: 63–78.

Box, E. O. 1981. *Macroclimate and Plant Forms: An Introduction to Predictive Modelling in Phytogeography.* Junk, The Hague.

Bradley, R. S., H. F. Diaz, J. K. Eischeid, P. D. Jones, P. M. Kelly and C. M. Goodess. 1987. Precipitation fluctuations over Northern Hemisphere land areas since the mid-nineteenth century. Science 237: 171–175.

Broecker, W. S. 1982. Ocean chemistry during glacial time. Geochem. Cosmochem. Acta 46: 1689–1705.

Broecker, W. S. and G. H. Denton. 1990. What drives glacial cycles? Sci. Am. 256: 49–56.

Bryson, R. A. 1966. Air masses, streamlines, and the boreal forest. Geograph. Bull. 8: 228–269.

Bryson, R. A. 1989. Late Quaternary volcanic modulation of Milankovitch climate forcing. Theor. Appl. Climatol. 39: 115–125.

Bryson, R. A. and T. J. Murray. 1977. *Climates of Hunger.* University of Wisconsin Press, Madison.

Buffington, L. C. and C. H. Herbel. 1965. Vegetational changes on a semidesert grassland range from 1858 to 1963. Ecol. Monogr. 35: 139–164.

Caldwell, M. M., R. Robberrecht, R. S. Nowak and W. D. Billings. 1982. Differential photosynthetic inhibition by ultraviolet radiation in species from the arctic–alpine life zone. Arc. Alp. Res. 14: 195–202.

Cannell, M. G. R., J. R. Grace and A. Booth. 1989. Possible impacts of climatic warming on trees and forests in the United Kingdom: A review. Forestry 62: 337–404.

Caprio, J. M. 1967. Phenological patterns and their use as climatic indicators. In *Ground Level Climatology.* American Association for the Advancement of Science, Washington, D.C.

Catchpole, A. J. W. and J. Hanuta. 1989. Severe summer ice in Hudson Strait and Hudson Bay following major volcanic eruptions, 1751 to 1889 A.D. Clim. Change 14: 61–79.

Chapin III, F. S., R. L. Jeffries, J. F. Reynolds, G. R. Shaver and J. Svoboda (eds.). 1992. *Arctic Ecosystems in a Changing Climate.* Academic Press, New York.

Chappellaz, J., J. M. Barnola, D. Raynaud, Y. S. Korotkevich and C. Lorius. 1990. Ice-core record of atmospheric methane over the past 160,000 years. Nature 345: 127–131.

Charlson, R. J., J. E. Lovelock, M. O. Andreae and S. G. Warren. 1987. Oceanic phytoplankton, atmospheric sulphur, cloud albedo, and climate. Nature 326: 655–661.

Charlson, R. J., S. E. Schwartz, J. M. Hales, R. D. Cess, J. A. Coakley Jr., J. E. Hansen and D. J. Hofmann. 1992. Climate forcing by anthropogenic aerosols. Science 255:

423–430.

Charney, J., P. H. Stone and W. J. Quirk. 1975. Drought in the Sahara: A biophysical feedback mechanism. Science 187: 434–435.

Clark, J. 1969. Thermal pollution and aquatic life. Sci. Am. 220: 19–27.

Clayton, H. H. 1923. *World Weather*. Macmillan, New York.

Clements, F. A. 1916. *Plant Succession: An Analysis of the Development of Species*. Carnegie Institution of Washington Publication No. 242. Reprinted 1928 by Wilson, New York.

CLIMAP Project Members. 1976. The surface of the ice-age Earth. Science 191: 1131–1136.

COHMAP Members. 1988. Climatic changes of the last 18,000 years: Observations and model simulation. Science 241: 1043–1052.

Coohill, T. P. 1989. Ultraviolet action spectra (280 to 380 nm) and solar effectiveness spectra for higher plants. Photochem. Photobiol. 50: 451–457.

Coope, G. R. 1977. Fossil coleopteran assemblages as sensitive indicators of climatic changes during the Devensian (Last) cold stage. Philos. Trans. R. Soc. Lond. B. 280: 313–340.

Craig, H., C. C. Chou, J. A. Welham, C. M. Stevens and A. Engelkemeir. 1988. The isotopic composition of methane in polar ice cores. Science 242: 1535–1539.

Crutzen, P. J. 1991. Methane's sinks and sources. Nature 350: 380–381.

Cullen, J. J., P. J. Neale and M. P. Lesser. 1992. Biological weighting function for the inhibition of phytoplankton photosynthesis by ultraviolet radiation. Science 258: 646–650.

Currie, R. G. 1981. Evidence for 18.6 yr signal in temperature and drought conditions in North America since 1800 A.D. J. Geophys. Res. 86: 11055–11064.

Dansgaard, W., J. W. C. White and S. J. Johnsen. 1989. The abrupt termination of the Younger Dryas climate event. Nature 339: 532–534.

Davis, M. B. 1981. Quaternary history and the stability of forest communities. In *Forest Succession: Concepts and Application*, ed. D. C. West, H. H. Shugart and D. B. Botkin. Springer-Verlag, New York. pp. 132–153.

Davis, M. B. 1987. Invasions of forest communities during the Holocene: Beech and hemlock in the Great Lakes region. In *Colonization, Succession and Stability*, ed. A. J. Gray, M. J. Crawley and P. J. Edwards. Blackwell Scientific, Oxford. pp. 373–393.

Davis, M. B. and D. B. Botkin. 1985. Sensitivity of cool-temperate forests and their fossil pollen record to rapid temperature change. Quat. Res. 23: 327–340.

Decker, W. L. 1983. The impacts of climatic variabilities on livestock production. In *Beef Cattle Science Handbook*, Vol. 19, ed. F. H. Baker. Westview Press, Boulder, CO. pp. 175–183.

Decker, W. L., V. K. Jones and R. Achutuni. 1986. The impact of climate change from increased atmospheric carbon dioxide on American agriculture. U.S. Department of Energy Report TR-031. Washington, D.C.

Delcourt, P. A. and H. R. Delcourt. 1981. Vegetation maps for eastern North America: 40,000 yr B.P. to the present. In *Geobotany II*, ed. R. C. Romans. Plenum Press, New York. pp. 123–165.

Delcourt, P. A. and H. R. Delcourt. 1983. Late Quaternary vegetational dynamics and community stability reconsidered. Quat. Res. 19: 265–271.

Diaz, H. F. 1983. Drought in the United States: Some aspects of major dry and wet periods in the contiguous United States, 1895–1981. J. Climatol. Appl. Meteorol. 22: 3–16.

Douglass, A. E. 1919. *Climatic Cycles and Tree Growth*, Vol. I. *A Study of the Annual Rings of Trees in Relation to Climate and Solar Activity*. Carnegie Institution of Washington Publication 289, Washington, D.C.

Douglass, A. E. 1928. *Climatic Cycles and Tree Growth*, Vol. II. *A Study of the Annual Rings of Trees in Relation to Climate and Solar Activity*. Carnegie Institution of Washington Publication 289, Washington, D.C.

Douglass, A. E. 1936. *Climatic Cycles and Tree Growth*, Vol. III. *A Study of Cycles*. Carnegie Institution of Washington Publication 289, Washington, D.C.

Dudek, D. J. 1989. Climate change impacts upon agriculture and resources: A case study of California. In Smith and Tirpak, 1989, Appendix C, pp. 5-1 to 5-38.

Dunham, A. E. 1993. Population responses to environmental change: Operative environments, physiologically structured models, and population dynamics. In *Biotic Interactions and Global Change*, ed. P. M. Kareiva, J. G. Kingsolver and R.

B. Huey. Sinauer Associates, Sunderland, MA. pp. 95–119.

Eddy, J. A. 1977. Climate and the changing sun. Clim. Change 1: 173–190.

Emanuel, W. R., H. H. Shugart and M. P. Stevenson. 1985a. Climatic change and the broad-scale distribution of terrestrial ecosystem complexes. Clim. Change 7: 29–43.

Emanuel, W. R., H. H. Shugart and M. P. Stevenson. 1985b. Response to comment: Climatic change and the broad-scale distribution of terrestrial ecosystem complexes. Clim. Change 7: 457–460.

Engstrom, D. R., B. C. S. Hansen and H. E. Wright Jr. 1990. A possible Younger Dryas record in southeastern Alaska. Science 250: 1383–1385.

Erkamo, V. 1952. On plant biological phenomena accompanying the present climatic change. Fennia 75: 25–37.

Fairbanks, R. G. 1989. A 17,000 year glacio-eustatic sea level record: Influence of glacial melting rates on the Younger Dryas event and deep-ocean circulation. Nature 342: 637–642.

Farman, J. C., B. G. Gardiner and J. D. Shanklin. 1975. Large losses of total ozone in Antarctica reveal seasonal ClO_x/NO_x interaction. Nature 315: 207–210.

Foukal, P. and J. Lean. 1990. An empirical model of total solar irradiance variation between 1874 and 1988. Science 247: 556–558.

Frederick, J. E. and H. E. Snell. 1988. Ultraviolet radiation levels during the Antarctic spring. Science 241: 438–440.

Friis-Christensen, E. and K. Lassen. 1991. Length of the solar cycle: An indicator of solar activity closely associated with climate. Science 254: 698–700.

Gates, D. M. 1972. *Man and His Environment: Climate.* Harper & Row, New York.

Gates, D. M. 1980. *Biophysical Ecology.* Springer-Verlag, New York.

Gates, D. M. 1985a. *Energy and Ecology.* Sinauer Associates, Sunderland, MA.

Gates, D. M. 1985b. Global biospheric response to increasing atmospheric carbon dioxide concentration. In *Direct Effects of Increasing Carbon Dioxide,* ed. B. R. Strain and J. D. Cure. U.S. Department of Energy DOE/ER-0238. Washington,

D.C. pp. 171–184.

Gates, D. M. 1990. Climate change and forests. Tree Physiol. 7: 1–5.

Gates, D. M. and L. E. Papian. 1971. *Atlas of Energy Budgets for Plant Leaves.* Academic Press, New York.

Gates, D. M., B. R. Strain and J. A. Weber. 1983. Ecophysiological effects of changing atmospheric CO_2 concentration. In *Encyclopedia of Plant Physiology,* New Series, Vol. 12D, *Physiological Plant Ecology IV,* ed. O. L. Lange, P. S. Nobel, C. B. Osmond and H. Ziegler. Springer-Verlag, New York.

Gates, W. L. 1976. Modeling the ice age climate. Science 191: 1138–1144.

Gear, A. J. and B. Huntley. 1991. Rapid changes in the range limits of Scots pine 4000 years ago. Science 251: 544–547.

George, M. F., M. J. Burke, H. M. Pellet and A. G. Johnson. 1974. Low temperature exotherms and woody plant distribution. Hort. Sci. 9: 519–522.

Graham, R. L., M. G. Turner and V. H. Dale. 1990. How increasing CO_2 and climate change affects forests. BioScience 40: 575–587.

Graham, R. W. 1992. Late Pleistocene faunal changes as a guide to understanding effects of greenhouse warming on the mammalian fauna of North America. In *Global Warming and Biological Diversity,* ed. R. L. Peters and T. E. Lovejoy. Yale University Press, New Haven. pp. 76–88.

Green, G. W. 1968. Weather and insects. In *Biometeorology,* ed. W. P. Lowry. Oregon State University Press, Corvallis. pp. 81–112.

Grootes, P. M. 1978. Carbon-14 time scale extended: Comparison of chronologies. Science 200: 11–15.

Grove, J. M. 1988. *The Little Ice Age.* Methuen, London.

Guetter, P. J. and J. E. Kutzbach. 1990. A modified Köppen classification applied to model simulations of glacial and interglacial climates. Clim. Change 16: 193–215.

Hamburg, S. P. and C. V. Cogbill. 1988. Historical decline of red spruce populations and climatic warming. Nature 331: 428–431.

Handler, P. and E. Handler. 1983. Climatic anomalies in the tropical Pacific Ocean and corn yields in the United States. Science 220: 1155–1156.

Hansen, J. and A. A. Lacis. 1990. Sun and dust versus greenhouse gases: An assessment of their relative roles in global climate change. Nature 346: 713–719.

Hansen, J. and S. Lebedeff. 1987. Global trends of measured surface air temperatures. J. Geophys. Res. 92: 13345–13372.

Hansen, J., I. Fung, A. Lacis, D. Rind, G. Russell, S. Lebedeff, R. Ruedy and P. Stone. 1988. Global climate changes as forecast by the GISS 3-D model. J. Geophys. Res. 93: 9341–9364.

Hansen, J., W. Rossow and I. Fung. 1990. The missing data on global climate change. In *Issues in Science and Technology*, Vol. 7. National Academy of Science, Washington, D.C. pp. 62–69.

Hanson, H. P., C. S. Hanson and B. H. Yoo. 1992. Recent Great Lakes ice trends. Bull. Am. Meteorol. Soc. 73: 577–584.

Hays, J. D., J. Imbrie and N. J. Shackleton. 1976. Variations in the Earth's orbit: Pacemaker of the ice ages. Science 194: 1121–1132.

Heusser, C. J., L. E. Heusser and D. M. Peteet. 1985. Late Quaternary North Pacific Coast. Nature 315: 485–487.

Hofmann, D. J. and T. Deshler. 1991. Evidence from balloon measurements for chemical depletion of stratospheric ozone in the Arctic winter of 1989–1990. Nature 349: 300–305.

Holdridge, L. R. 1947. Determinations of world formulations from simple climatic data. Science 105: 367–368.

Hopkins, A. D. 1918. Periodical events and natural laws as guides to agricultural research and practice. U.S. Department of Agriculture Monthly Weather Review, Supplement 9.

Hulbert, L. C. 1963. Gates' phenological records of 132 plants at Manhattan, Kansas, 1926–1955. Trans. Kans. Acad. Sci. 66: 82–106.

Hunt, H. W., M. J. Trlica, E. F. Redente, J. C. Moore, J. K. Detling, T. G. F. Kittel, D. E. Walter, M. C. Fowler, D. A. Klein and E. T. Elliott. 1991. Simulation model for the effects of climate change on temperate grassland ecosystems. Ecol. Modelling 53: 205–246.

Hustich, I. 1952. The recent climatic fluctuation in Finland and its consequences. Fennia 75: 1–28.

Huston, M. and T. M. Smith. 1987. Plant succession: Life history and competition.

Am. Nat. 130: 168–198.

Imbrie, J. and J. Z. Imbrie. 1980. Modeling the climatic response to orbital variations. Science 207: 943–953.

Jacoby, G. C. and R. D'Arrigo. 1989. Reconstructed Northern Hemisphere temperature since 1671 based on high-altitude tree-ring data from North America. Clim. Change 14: 39–59.

Jacoby, G. C., I. S. Ivanciu and L. D. Ulan. 1988. A 263-year record of summer temperature for northern Quebec reconstructed from tree-ring data and evidence of a major climatic shift in the early 1800s. Palaeogeog. Palaeoclimatol. Palaeoecol. 64: 69–78.

Jones, C. A. and J. R. Kiniry (eds.) 1986. CERES-Maize: A simulation model of maize growth and development. Texas A&M University Press, College Station, TX.

Jones, J. W., T. W. Mishoe, G. Wilkerson, J. L. Stimac and W. G. Boggess. 1986. Integration of soybean crop and pest models. In *Integrated Pest Management on Major Agricultural Systems*, ed. R. E. Frisbie and P. E. Adhisson. Texas Agricultural Experiment Station Misc. Publ. MP-1616.

Jones, P. D. 1988. Hemispheric surface-air temperature variations: Recent trends and an update to 1987. J. Climatol. 1: 654–660.

Jones, P. D. and T. M. L. Wigley. 1990. Global warming trends. Sci. Am. 263: 84–91.

Jones, P. D. and T. M. L. Wigley. 1991. The global temperature record for 1990. Department of Energy Research Summary No. 10. Carbon Dioxide Information Analysis Center, Oak Ridge National Laboratory, Oak Ridge, TN.

Jones, R. L. 1989. Ozone chemistry depletion on volcanic aerosols. Nature 340: 269–270.

Joos, F., J. L. Sarmiento and U. Siegenthaler. 1991. Estimates of the effect of Southern Ocean iron fertilization on atmospheric CO_2 concentrations. Nature 349: 772–775.

Karl, T. R., J. D. Tarpley, R. G. Quayle, H. F. Diaz, D. A. Robinson and R. S. Bradley. 1989. The recent climate record: What it can and cannot tell us. Rev. Geophys. 27:

405–430.

Karl, T. R., G. Kukla, V. N. Razuvayev, M. J. Changery, R. G. Quayle, R. R. Heim, Jr., D. R. Easterling and C. B. Fu. 1991. Global warming: Evidence for asymmetric diurnal temperature change. Geophys. Res. Lett. 18: 2253–2256.

Karentz, S., J. E. Cleaver and D. L. Mitchell. 1991. DNA damage in the Antarctic. Nature 350: 28.

Kauppi, P. and M. Posch. 1985. Sensitivity of boreal forests to possible climatic warming. Clim. Change 7: 45–54.

Kauppi, P. and M. Posch. 1988. A case study of the effects of CO_2-induced climatic warming on forest growth and the forest sector. A. Productivity reactions of northern boreal forests. In *The Impact of Climatic Variations on Agriculture*, Vol. I, *Assessments in Cool Temperate and Cold Regions*, ed. M. L. Parry, T. R. Carter and N. T. Konijn. Kluwer Academic, Dordrecht, The Netherlands. pp. 183 - 195.

Kay, P. A. 1979. Multivariate statistical estimates of Holocene vegetation and climate change, forest–tundra transition zone, Northwest Territories, Canada. Quat. Res. 11: 125–140.

Kellogg, W. W. 1991. Response to skeptics of global warming. Bull. Am. Meteorol. Soc. 74: 499–511.

Kelly, M. 1990. Halting global warming. In *Global Warming: The Greenpeace Report*, ed. J. Leggett. Oxford University Press, New York.

Kelly, P. M. and C. B. Sear. 1984. Climatic impact of explosive volcanic eruptions. Nature 311: 740–743.

Kerr, R. A. 1990. Global warming continues in 1989. Science 247: 521.

King, J. W. 1973. Solar radiation changes and the weather. Nature 245: 443.

Klein, R. M. and T. D. Perkins. 1988. Primary and secondary causes and consequences of contemporary forest decline. Bot. Rev. 54: 1–43.

Klinger, L. F. 1991. Peatland formation and ice ages: A possible Gaian mechanism related to community succession. In *Scientists on Gaia*. S. H. Schneider and P. J. Boston (eds.) M.I.T. Press, Cambridge, MA, pp. 247–255.

Kudrass, H. R., H. Erienkeuser, R. Vollbrecht and W. Weiss. 1991. Global nature of the Younger Dryas cooling event inferred from oxygen isotope data from Sulu Sea cores. Nature 349: 406–409.

Kullman, L. 1979. Change and stability in the altitude of the birch tree limit in the southern Swedish Scandes 1915–1975. Acta Phytogeographica Suecica 65: 1–121.

Kutzbach, J. E. 1985. Modeling of paleoclimates. Adv. Geophys. 28A: 159–196.

Labitzke, K. and H. van Loon. 1988. Association between the 11-year solar cycle, the QBO and the atmosphere. I. The troposphere and stratosphere in the Northern Hemisphere in winter. J. Atmos. Terr. Phys. 50: 197–206.

Ladurie, E. L. 1971. *Times of Feast, Times of Famine: A History of Climate since the Year 1000*. Garden City Press, New York.

Lal, M. and T. Holt. 1991. Ozone depletion due to increasing anthropogenic trace gas emissions: Role of stratospheric chemistry and implications for future climate. Clim. Res. 1: 85–95.

Lamb, H. H. 1970. Volcanic dust in the atmosphere; with a chronology and assessment of its meteorological significance. Philos. Trans. R. Soc. Lond. A 266: 425–533.

Lamb, H. H. 1977. *Climate: Present, Past and Future*, Vol. 2. Methuen, London.

Lang, A. 1965. Effects of some internal and external conditions on seed germination. In *Encyclopedia of Plant Physiology*, Vol. 15, ed. W. Ruhland. Springer-Verlag, New York. pp. 848–893.

Larcher, W. 1975. *Physiological Plant Ecology*. Translated and revised from the German. Springer-Verlag, New York.

Larcher, W. 1980. *Physiological Plant Ecology*, 2nd Ed. Springer-Verlag, New York.

Leaphart, C. D. and A. R. Stage. 1971. Climate: A factor in the origin of the pole blight disease of *Pinus monticola* Dougl. Ecology 52: 229–239.

Leggett, J. (ed.). 1990. *Global Warming: The Greenpeace Report*. Oxford University Press, New York.

Lemon, E. R. (ed.). 1984. CO_2 *and Plants: The Response of Plants to Rising Levels of Atmospheric Carbon Dioxide*. Westview Press, Boulder, CO.

Lettenmaier, D. P., T. Y. Gan and D. R. Dawdy. 1989. Interpretation of hydrologic effects of climate change in the Sacramento–San Joaquin River Basin, California. In Smith and Tirpak, 1989,

Appendix A, pp. 1-1 to 1-52.

Liss, P. S. and A. J. Crane. 1983. *Man-Made Carbon Dioxide and Climate Change: A Review of Scientific Problems*. GEO Books, Norwich, England.

Lorius, C., J. Jauzel, D. Raynaud, J. Hansen and H. LeTreut. 1990. The ice-core record: Climate sensitivity and future greenhouse warming. Nature 347: 139–145.

Lough, J. M. and H. C. Fritts. 1987. An assessment of the possible effects of volcanic eruptions on North American climate using tree-ring data, 1602 to 1900 A.D. Clim. Change 10: 219–239.

Lovelock, J. E. 1979. *Gaia: A New Look at Life on Earth*. Oxford University Press, Oxford.

Magnuson, J. J., J. D. Meisner and D. K. Hill. 1990. Potential changes in the thermal habitat of Great Lakes fish after climate warming. Trans. Am. Fish. Soc. 119: 254–264.

Manabe, S. and R. J. Stouffer. 1980. Sensitivity of a global climate model to an increase of CO_2 concentration in the atmosphere. J. Geophys. Res. 85: 5529–5554.

Manabe, S. and R. T. Wetherald. 1987. Large scale changes in soil wetness induced by an increase in carbon dioxide. J. Atmos. Sci. 44: 1211–1235.

Manabe, S., R. T. Wetherald and R. J. Stouffer. 1981. Summer dryness due to an increase of atmospheric CO_2 concentration. Clim. Change 3: 347–386.

Mass, C. F. and D. A. Portman. 1989. Major volcanic eruptions and climate: A critical evaluation. J. Clim. 2: 566–593.

McCormick, J. H., K. M. Jensen and R. L. Leino. 1990. Survival, blood osmolality and gill morphology of juvenile yellow perch, rock bass, black crappie and largemouth bass exposed to acidified soft water. Trans. Am. Fish. Soc. 118: 386–399.

McElroy, M. B. and R. J. Salawitch. 1989. Changing composition of the global stratosphere. Science 243: 763–770.

Mearns, L. O., R. W. Katz and S. H. Schneider. 1984. Extreme high temperature events: changes in their probabilities and changes in mean temperature. J. Climatol. Appl. Meteor. 23: 1601–1613.

Messenger, P. S. 1959. Bioclimatic studies with insects. Annu. Rev. Entomol. 4: 183–206.

Michaels, P. J. and D. E. Stooksbury. 1992. Global warming: A reduced threat. Bull. Am. Meteorol. Soc. 73: 1563–1577.

Miller, J. A. 1986. Computer modeling populations of the horn fly. In *Modeling and Simulation: Tools for Management of Veterinary Pests*, ed. J. A. Miller. U.S. Department of Agriculture ARS-46. pp. 33–40.

Miller, J. A. 1989. Research update: Goose prelude to global warming. BioScience 39: 673.

Mitchell, J. F. B. 1983. The seasonal response of a general circulation model to changes in CO_2 and sea temperatures. Q. J. R. Meteorol. Soc. 109: 113–152.

Mitchell, J. F. B. and G. Lupton. 1984. A 4 x CO_2 integration with prescribed changes in sea surface temperatures. Prog. Biometeorol. 3: 353–374.

Mitchell, J. M. 1982. El Chichón: Weather-maker of the century? Weatherwise 35: 252–262.

Mitchell, J. M. Jr., C. W. Stockton and D. M. Meko. 1979. Evidence of a 22-year rhythm of drought in the western United States related to the Hale Solar Cycle since the seventeenth century. In *Solar and Terrestrial Influence on Weather and Climate*, ed. B. M. McCormac and T. A. Seliga. D. Reidel, Dordrecht, The Netherlands. pp. 155–174.

Namias, J. 1989. Cold waters and hot summers. Nature 338: 15–16.

National Research Council, Geophysics Study Committee. 1982. *Studies in Geophysics: Solar Variability, Weather and Climate*. National Academy Press, Washington, D.C.

Neftel, A., H. Oeschger, J. Schwander, B. Stauffer and R. Zumbrunn. 1982. Ice core sample measurements give atmospheric CO_2 content during the past 40,000 years. Nature 295: 220–223.

Newman, J. E. 1980. Climate change impacts on the growing season of the North American corn belt. Biometeorology 7: 128–142.

Nichols, H. 1975. Palynological and paleoclimatic study of the late Quaternary displacement of the boreal forest-tundra ecotone in Keewatin and MacKenzie, N.W.T., Canada. University of Colorado Institute for Arctic and Alpine Research, Occ. Pap. 15.

Nisbet, E. 1990. Climate change and methane. Nature 347: 23.

Odum, E. P. 1993. *Ecology and Our Endangered Life-Support Systems*, 2nd Ed. Sinauer Associates, Sunderland, MA.

Okamoto, K., T. Okiwara, T. Yoshizumi and Y. Watanabe. 1991. Influence of the greenhouse effect on yields of wheat, soybean and corn in the United States for different energy scenarios. Clim. Change 18: 397–424.

Overpeck, J. T. and P. J. Bartlein. 1989. Assessing the response of vegetation to future climate change: ecological response surfaces and paleoecological model validation. In Smith and Tirpak, 1989, Appendix D, pp. 1-1 to 1-32.

Overpeck, J. T., L. C. Petterson, N. Kipp, J. Imbrie and D. Rind. 1989. Climate change in the circum-North Atlantic region during the last deglaciation. Nature 338: 553–557.

Overpeck, J. T., P. J. Bartlein and T. Webb. 1991. Potential magnitude of future vegetation change in eastern North America: Comparisons with the past. Science 254: 692–695.

Pastor, J. and W. M. Post. 1986. Influence of climate, soil, moisture and succession on forest carbon and nitrogen cycles. Biogeochemistry 2: 3–27.

Pastor, J. and W. M. Post. 1988. Response of northern forests to CO_2 induced climate change. Nature 334: 55–58.

Peart, R. M., J. W. Jones, R. B. Curry, K. Boote and L. H. Allen. 1989. Impact of climate change on crop yield in the southeastern U.S.A.: A simulation study. In Smith and Tirpak, 1989, Appendix C, pp. 2-1 to 2-54.

Peltier, W. R. and A. M. Tushingham. 1989. Global sea level rise and the greenhouse effect: Might they be connected? Science 244: 806–810.

Peteet, D. M., J. S. Vogel, D. E. Nelson, J. R. Southon, R. J. Nickmann and L. E. Heusser. 1990. Younger Dryas climatic reversal in northeastern U.S.A.? AMS ages for an old problem. Quat. Res. 33: 219–230.

Peters, R. L. and T. E. Lovejoy (eds.). 1992. *Global Warming and Biological Diversity*. Yale University Press, New Haven.

Philander, G. 1989. *El Niño, La Niña and the Southern Oscillation*. Academic Press, San Diego, CA.

Pigott, C. D. 1981. Nature of seed sterility and natural regeneration of *Tilia cordata* near its northern limit in Finland. Ann. Bot. Fennici 18: 255–263.

Pigott, C. D. and J. P. Huntley. 1981. Factors controlling the distribution of *Tilia cordata* at the northern limits of its geographical range. III. Nature and causes of seed sterility. New Phytol. 87: 817–839.

Pisias, N. G. and N. J. Shackleton. 1984. Modelling the global climate response to orbital forcing and atmospheric carbon dioxide changes. Nature 310: 757–759.

Plantico, M. S., T. R. Karl, G. Kukla and J. Gavin. 1990. Is the recent climate change across the United States related to rising levels of anthropogenic greenhouse gases? J. Geophys. Res. 95: 16617–16637.

Poiani, K. A. and W. C. Johnson. 1991. Global warming and prairie wetlands. BioScience 41: 611–618.

Pomerleau, R. and M. Lortie. 1962. Relationships of dieback to the rooting depth of white birch. For. Sci. 8: 219–224.

Porter, J. H., M. L. Parry and T. R. Carter. 1991. The potential effects of climatic change on agricultural insect pests. Agric. For. Meteorol. 57: 221–240.

Porter, W. P. and D. M. Gates. 1969. Thermodynamic equilibria of animals with the environment. Ecol. Monogr. 39: 245–270.

Pounds, A. 1991. Costa Rican frog and toad populations indicative of worldwide amphibian response to climate change. UPDATE 5: 3–4. Newsletter of the Center for Conservation Biology, Stanford University, Stanford, CA.

Prell, W. L. and J. E. Kutzbach. 1987. Monsoon variability over the past 150,000 years. J. Geophys. Res. 92: 8411–8425.

Prentice, I. C., R. S. Webb, M. T. Ter-Mikhaelian, A. M. Solomon, T. M. Smith, S. E. Pitovranov, N. T. Nikolov, A. A. Minin, R. Leemans, S. Lavorel, M. D. Korzukhin, J. P. Hrabovsky, H. O. Helmisaari, S. P. Harrison, W. R. Emanuel and G. B. Bonan. 1989. Developing a Global Vegetation Dynamics Model: Results of an IIASA Summer Workshop. International Institute for Applied Systems Analysis, Laxenburg, Austria.

Prentice, I. C., M. T. Sykes and W. Cramer.

1991. The possible dynamic response of northern forests to global warming. Global Ecol. Biogeog. Lett. 1: 129–135.

Prentice, I. C., P. J. Bartlein and T. Webb. 1992a. Vegetation and climate change in eastern North America since the glacial maximum. Ecology 72: 2038–2056.

Prentice, I. C., W. Cramer, S. P. Harrison, R. Leemans, R. A. Monserud and A. M. Solomon. 1992b. A global biome model based on plant physiology and dominance, soil properties, and climate. Biogeography, in press.

Prentice, K. C. 1990. Bioclimatic distribution of vegetation for general circulation model studies. J. Geophys. Res. 95: 11811–11830.

Rasmussen, E. M. 1985. El Niño and variations in climate. Am. Sci. 73: 168–177.

Rasmussen, E. M., X. Wang and C. F. Ropelewski. 1990. The biennial component of ENSO variability. J. Mar. Syst. 1: 71–96.

Raval, A. and V. Ramanathan. 1989. Observational determination of the greenhouse effect. Nature 342: 758–761.

Rhoades, D. F. 1985. Offensive-defensive interactions between herbivores and plants: Their relevance in herbivore population dynamics and ecological theory. Am. Nat. 125: 205–238.

Rind, D., D. Peteet, W. Broecker, A. McIntyre and W. Ruddiman. 1986. The impact of cold North Atlantic sea surface temperatures on climate: implications for the Younger Dryas cooling (11–10 kya). Clim. Dynam. 1: 3–33.

Rind, D., E. W. Chiou, W. Chu, J. Larsen, S. Oltmans, J. Lerner, M. P. McCormick and L. McMaster. 1991. Positive water vapour feedback in climate models confirmed by satellite data. Nature 349: 500–503.

Ritchie, J. T. and S. Otter. 1985. Description and performance of CERES-Wheat: A user-oriented wheat yield model. In *ARS Wheat Yield Project*, ed. W. O. Willis, U.S. Department of Agriculture ARS-38. pp. 159–175.

Ritchie, J. T., B. D. Baer and T. Y. Chou. 1989. Effect of global change on agriculture Great Lakes region. In Smith and Tirpak, 1989, Appendix C, pp. 1-1 to 1-30.

Robock, A. 1979. The "Little Ice Age":

Northern Hemisphere average observations and model calculations. Science 206: 1402–1404.

Rodhe, H. 1990. A comparison of the contribution of various gases to the greenhouse effect. Science 248: 1217–1219.

Root, T. 1988a. Environmental factors associated with avian distributional boundaries. J. Biogeogr. 15: 489–505.

Root, T. 1988b. Energy constraints on avian distributions and abundances. Ecology 69: 330–339.

Rosenberg, N. J. 1991. *Processes for Identifying Regional Influences of and Responses to Increasing Atmospheric CO$_2$ and Climate Change: The MINK Project.* Report I, Background and Baselines. Resources for the Future, Washington, D.C.

Rosenzweig, C. 1985. Potential CO$_2$-induced climate effects on North American wheat-production regions. Clim. Change 7: 367–389.

Rosenzweig, C. 1989. Potential effects of climate change on agricultural production in the Great Plains: A simulation study. In Smith and Tirpak, 1989, Appendix C, pp. 3-1 to 3-43.

Rosenzweig, C. 1990. Crop response to climate change in the southern Great Plains: A simulation study. Prof. Geogr. 42: 20–37.

Rowntree, P. R. 1985. Comment on "Climatic change and the broad-scale distribution of terrestrial ecosystem complexes" by Emanuel, Shugart and Stevenson. Clim. Change 7: 455–456.

Sargent, N. E. 1988. Redistribution of the Canadian boreal forest under a warmed climate. Clim. Bull. 22: 23–34.

Schindler, D. W., K. G. Beaty, E. J. Fee, D. R. Cruikshank, E. R. DeBruyn, D. L. Findlay, G. A. Linsey, J. A. Shearer, M. P. Stainton and M. A. Turner. 1990. Effects of climatic warming on lakes of the central boreal forest. Science 250: 967–970.

Schlesinger, M. and Z. Zhao. 1988. Seasonal climate changes induced by doubled CO$_2$ as simulated by the OSU atmospheric GCM/mixed-layer ocean model. Climate Research Institute, Oregon State University, Corvallis.

Schlesinger, W. H., J. F. Reynolds, G. L. Cunningham, L. F. Huenneke, W. M. Jarrell, R. A. Virginia and W. B.

Whitford. 1990. Biological feedbacks in global desertification. Science 247: 1043–1048.

Schmidtmann, E. T. and J. A. Miller. 1989. Effect of climate warming on populations of the horn fly, with associated impact on weight gain and milk production in cattle. In Smith and Tirpak, 1989, Appendix C, pp. 12-1 to 12-11.

Schneider, S. H. 1990. The global warming debate heats up: An analysis and perspective. Bull. Amer. Meteorol. Soc. 71: 1291–1304.

Schreiber, R. W. and E. A. Schreiber. 1984. Central Pacific seabirds and the El Niño southern oscillation: 1982 to 1983 perspective. Science 225: 713–716.

Schwartz, M. D. 1990. Detecting the onset of spring: A possible application of phenological models. Clim. Res. 1: 23–29.

Seitz, F., R. Jastrow and W. A. Nierenberg. 1989. *Scientific Perspectives on the Greenhouse Problem.* George C. Marshall Institute, Washington, D.C.

Shackleton, N. J. 1989. Deep trouble for climate change. Nature 342: 616–617.

Shaver, G. R., W. D. Billings, F. Chapin III, A. E. Giblin, K. J. Nadelhoffer, W. C. Oechel and E. B. Rastetter. 1992. Global change and the carbon balance of arctic ecosystems. BioScience 42: 433–441.

Shaw, N. 1928. *Manual of Meteorology*, Vol. II, *Comparative Meteorology.* Cambridge University Press, Cambridge.

Shugart, H. H. and D. C. West. 1977. Development of an Appalachian deciduous forest succession model and its application to assessment of the impact of the chestnut blight. J. Environ. Manage. 5: 161–179.

Smith, J. B. 1991. The potential impacts of climate change on the Great Lakes. Bull. Am. Meteorol. Soc. 72: 21–28.

Smith, J. B. and D. Tirpak. 1989. *The Potential Effects of Global Climate Change on the United States.* U.S. Environmental Protection Agency, Washington, D.C.

Smith, R. C., B. B. Prezelin, K. S. Baker, R. R. Bidigare, N. P. Boucher, T. Coley, D. Karentz, S. MacIntyre, H. A. Matlick, D. Menzies, M. Ondrusek, Z. Wan and K. J. Waters. 1992. Ozone depletion: Ultraviolet radiation and phytoplankton biology in Antarctic waters. Science 255: 952–958.

Solomon, A. M. 1986. Transient response of forests to CO_2-induced climate change: Simulation modeling experiments in eastern North America. Oecologia 68: 567–579.

Solomon, A. M. and P. J. Bartlein. 1992. Past and future climate change: Response by mixed deciduous–coniferous forest ecosystems in northern Michigan. Canad. J. For. Res., in press.

Solomon, A. M. and T. Webb. 1985. Computer-aided reconstruction of late Quaternary landscape dynamics. Annu. Rev. Ecol. Syst. 16: 63–84.

Solomon, A. M., D. C. West and J. A. Solomon. 1981. Simulating the role of climate change and species immigration in forest succession. In *Forest Succession: Concepts and Applications,* ed. D. C. West, H. H. Shugart and D. B. Botkin. Springer-Verlag, New York. pp. 154–177.

Solomon, A. M., M. L. Thorp, D. C. West, G. E. Taylor, J. W. Webb and J. L. Trimble. 1984. Response of unmanaged forests to CO_2-induced climate change: Available information, initial tests, and data requirements. U.S. Department of Energy Report TR-009. Washington, D.C.

Sperry, J. S. and M. T. Tyree. 1988. Mechanism of water stress-induced xylem embolism. Plant Physiol. 88: 581–587.

Spotila, J. R. 1972. Role of temperature and water in the ecology of lungless salamanders. Ecol. Monogr. 42: 95–125.

Staley, J. M. 1965. Decline and mortality of red and scarlet oaks. For. Sci. 11: 1–17.

Stem, E., G. A. Mertz, J. D. Stryker and M. Huppi. 1989. Changing animal disease patterns induced by the greenhouse effect. In Smith and Tirpak, 1989, Appendix C, pp. 11-1 to 11-37.

Stephenson, N. L. 1990. Climate control of vegetation distribution: The role of water balance. Am. Nat. 135: 649–670.

Stinner, B. R., R. A. J. Taylor, R. B. Hammond, F. F. Purrington, D. A. McCartney, N. Rodenhouse and G. W. Barrett. 1989. Potential effects of climate change on plant–pest interactions. In Smith and Tirpak, 1989, Appendix C, pp. 8-1 to 8-35.

Stommel, H. and E. Stommel. 1983. *Volcano Weather: The Story of 1816, the Year without a Summer.* Seven Seas Press, Newport, RI.

Stouffer, R. J., S. Manabe and K. Bryan. 1989. Interhemispheric asymmetry in climate response to a gradual increase of atmospheric CO_2. Nature 342: 660–662.

Strong, A. E. 1989. Greater global warming revealed by satellite derived sea-surface temperature trends. Nature 338: 642–645.

Strothers, R. B. 1984. The great Tambora eruption in 1815 and its aftermath. Science 224: 1191–1198.

Stuiver, M., C. J. Heusser and I. C. Yang. 1978. North American glacial history extended to 75,000 years ago. Science 200: 16–21.

Tans, P. P., I. Y. Fung and T. Takahashi. 1990. Observational constraints on the global atmospheric CO_2 budget. Science 247: 1431–1438.

Thompson, L. G., E. Mosley-Thompson, W. Dansgaard and P. M. Grootes. 1986. The Little Ice Age as recorded in the stratigraphy of the tropical Quelccaya ice cap. Science 234: 361–364.

Titus, J. G. (ed.). 1988. *Greenhouse Effect, Sea Level Rise and Coastal Wetlands.* U.S. Environmental Protection Agency, Washington, D.C.

Trenberth, K. E. 1990. Recent observed interdecadal climate changes in the Northern Hemisphere. Bull. Am. Meteorol. Soc. 71: 988–994.

Trenberth, K. E., G. W. Branstator and P. A. Arkin. 1988. Origins of the 1988 North American drought. Science 242: 1640–1645.

Tsonis, A. A. and J. B. Eisner. 1989. Testing the global warming hypothesis. Geophys. Res. Lett. 16: 795–797.

Tyree, M. T. and J. S. Sperry. 1988. Do woody plants operate near the point of catastrophic xylem dysfunction caused by dynamic water stress? Plant Physiol. 88: 574–580.

Urbach, F. (ed.). 1969. *The Biological Effects of Ultraviolet Radiation.* Pergamon, New York.

Urban, D. L. and H. H. Shugart. 1989. Forest response to climatic change: A simula-tion study for southeastern forests. In Smith and Tirpak, 1989, Appendix D, pp. 3-1 to 3-45.

Uvarov, B. P. 1931. Insects and climate. Trans. Entomol. Soc. Lond. 79: 1–247.

Van Cleve, K., F. S. Chapin III, C. T. Dryness and L. A. Viereck. 1991. Element cycling in taiga forests: State factor control.

BioScience 41: 78 -88.

Viereck, L. A. and K. Van Cleve. 1984. Some aspects of vegetation and temperature relationships in the Alaska taiga. In *The Potential Effects of Carbon Dioxide-Induced Climatic Changes in Alaska,* ed. J. H. McBeath. Misc. Publ. 83-1, School of Agriculture and Land Resources Management, University of Alaska, Fairbanks. pp. 129–142.

Voytek, M. A. 1990. Addressing the biological effects of decreased ozone on the Antarctic environment. Ambio 19: 52–61.

Wahlen, M., N. Tanaka, R. Henry, B. Deck, J. Zeglen, J. S. Vogel, J. Southon, A. Shemesh, R. Fairbanks and W. Broecker. 1989. Carbon-14 in methane sources and in atmospheric methane: The contribu-tion from fossil carbon. Science 245: 286–290.

Walker, B. H. 1991. Ecological consequences of atmospheric and climate change. Clim. Change 18: 301–316.

Walsh, J. E. 1991. The arctic as a bellwether. Nature 352: 19–20.

Washington, W. M. 1990. Where's the heat? Nat. Hist. 3: 67–72.

Washington, W. M. and G. A. Meehl. 1989. Climate sensitivity due to increased CO_2: Experiments with a coupled atmosphere and ocean general circulation model. Clim. Dynam. 4: 1–38.

Watts, W. A. 1979. Late Quaternary vegetation of central Appalachia and the New Jersey coastal plain. Ecol. Monogr. 49: 427–469.

Webb, T. III. 1986. Is vegetation in equilib-rium with climate? How to interpret late Quaternary pollen data. Vegetatio 67: 75–91.

Webb, T. III. 1987. The appearance and disappearance of major vegetational assemblages: Long-term vegetational dynamics in eastern North America. Vegetatio 69: 177–187.

Webb, T., P. J. Bartlein and J. E. Kutzbach. 1987. Climatic change in eastern North America during the past 18,000 years: Comparisons of pollen data with model results. In *North America and Adjacent Oceans during the Last Deglaciation,* Vol. K3, ed. W. F. Ruddiman and H. E. Wright Jr. Geological Society of America, Boulder, CO. pp. 447–462.

Went, F. W. 1957. *The Experimental Control of*

Plant Growth. Ronald Press, New York.

Went, F. W. 1961. Temperature: Survey, thermoperiodicity. In *Encyclopedia of Plant Physiology*, Vol. 16, ed. W. Ruhland. Springer-Verlag, Berlin. pp. 1 -23.

Wigley, T. M. L. and S. C. B. Raper. 1987. Thermal expansion of sea water associated with global warming. Nature 247: 127–131.

Wigley, T. M. L., P. D. Jones and P. M. Kelly. 1980. Scenario for a warm, high-CO_2 world. Nature 283: 17–21.

Wilkerson, G. G., J. W. Jones, K. J. Boote, K. T. Ingram and J. W. Mishoe. 1983. Modeling soybean growth for management. Trans. Am. Soc. Agric. Eng. 26: 63–73.

Woodward, F. I. 1987a. *Climate and Plant Distribution*. Cambridge University Press, Cambridge.

Woodward, F. I. 1987b. Stomatal numbers are sensitive to increases in CO_2 from pre-industrial levels. Nature 327: 617–618.

Woodward, F. I. 1992. A review of the effects of climate on vegetation: Ranges, competition, and composition. In *Global Warming and Biological Diversity*, ed. R. L. Peters and T. E. Lovejoy. Yale University Press, New Haven. pp. 105–123.

Wuebbles, D. J. and J. Edmonds. 1988. A primer on greenhouse gases. U.S. Department of Energy Report TR-040, Washington, D.C.

Zabinski, C. and M. B. Davis. 1989. Hard times ahead for Great Lakes forests: A climate threshold model predicts responses to CO_2-induced climate change. In Smith and Tirpak, 1989, Appendix D, pp. 5-1 to 5-19.

Zwally, H. J. 1989. Growth of Greenland Ice Sheet: Interpretation. Science 246: 1589–1591.

Index